Tethers In Space Handbook

Edited by
M.L. Cosmo and E.C. Lorenzini
Smithsonian Astrophysical Observatory

for
NASA Marshall Space Flight Center
Grant NAG8-1160 monitored by C.C. Rupp
M.L. Cosmo and E.C. Lorenzini, Principal Investigators

Third Edition
December 1997

The Smithsonian Astrophysical Observatory
is a member of the
Harvard-Smithsonian Center for Astrophysics

FOREWORD

A new edition of the *Tethers in Space Handbook* was needed after the last edition published in 1989. Tether-related activities have been quite busy in the 90's. We have had the flights of TSS1 and TSS1-R, SEDS-1 and -2, PMG, TIPS and OEDIPUS. In less than three years there have been one international Conference on Tethers in Space, held in Washington DC, and three workshops, held at ESA/Estec in the Netherlands, at ISAS in Japan and at the University of Michigan, Ann Harbor. The community has grown and we finally have real flight data to compare our models with. The life of spaceborne tethers has not been always easy and we got our dose of setbacks, but we feel pretty optimistic for the future. We are just stepping out of the pioneering stage to start to use tethers for space science and technological applications. As we are writing this handbook TiPs, a NRL tether project is flying above our heads.

There is no emphasis in affirming that as of today spaceborne tethers are a reality and their potential is far from being fully appreciated. Consequently, a large amount of new information had to be incorporated into this new edition.

The general structure of the handbook has been left mostly unchanged. The past editors have set a style which we have not felt needed change. The section on the flights has been enriched with information on the scientific results. The categories of the applications have not been modified, and in some cases we have mentioned the existence of related flight data.

We felt that the section contributed by Joe Carroll, called *Tether Data*, should be maintained as it was, being a "classic" and still very accurate and not at all obsolete.

We have introduced a new chapter entitled *Space Science and Tethers* since flight experience has shown that tethers can complement other space-based investigations.

The bibliography has been updated. Due to the great production in the last few years we had to restrict our search to works published in refereed journal. The production, however, is much more extensive. In addition, we have included the summary of the papers presented at the last International Conference which was a forum for first-hand information on all the flights.

We would like to thank the previous editors, W. Baracat and C. Butner, P.Penzo and P. Amman, for having done such a good job in the past editions that has made ours much easier.

The completion of this handbook would not have been possible without the contributions from the following people:

A. Allasio	A. Jablonski	J. Puig-Suari
F. Angrilli	L. Johnson	W. Purdy
S. Bergamaschi	K. Kirby	C. Rupp
M. Candidi	J. Longuski	D. Sabath
J. Carroll	M. Martinez-Sanchez	J. Sanmartin
K. Chance	P. Merlina	A. Santangelo
S. Coffey	L. Minna	S. Sasaki
D. Crouch	J. McCoy	N. Stone
R. Estes	A. Misra	B. Strim
L. Gentile	V. Modi	T. Stuart
F. Giani	P. Musi	G. Tacconi
M. Grossi	M. Novara	G. Tyc
D. Hardy	K. Oyama	F. Vigneron
R. Hoyt	P. Penzo	M. Zedd

Also, we would like to thank the staff of the Science Media Group at SAO for their help. NASA support for this work through Grant NAS8-1160 from NASA Marshall Space Flight Center is gratefully acknowledged.

Mario L. Cosmo
Enrico C. Lorenzini

Smithsonian Astrophysical Observatory
Cambridge, Massachusetts

December 1997

Tethers in Space Handbook - Third Edition
Table of Contents

SECTION 5.0 TETHER DATA

SECTION 6.0 SPACE SCIENCE AND TETHERS

SECTION 7.0 REFERENCES

SECTION 8.0 CONTACTS

SECTION 1.0 TETHER FLIGHTS

1.1 The Tethered Satellite System Program: TSS-1 and TSS-1R Missions

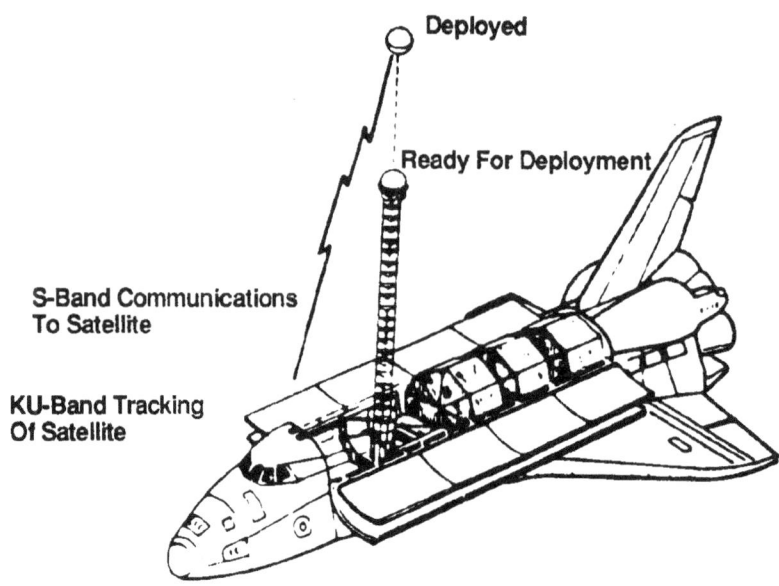

Figure 1.1 TSS-1 Satellite and Tether Attached to 12 Meter Extendible Boom

The Tethered Satellite System (TSS) was proposed to NASA and the Italian Space Agency (ASI) in the early 1970's by Mario Grossi, of the Smithsonian Astrophysical Observatory, and Giuseppe Colombo, of Padua University. A science committee, the Facilities Requirements Definition Team (FRDT), met in 1979 to consider the possible scientific applications for long tethers in space and whether the development of a tethered system was justified. The FRDT report, published in 1980, strongly endorsed a Shuttle-based tether system. A NASA-ASI memorandum of understanding was signed in 1984, in which NASA agreed to develop a deployer system and tether and ASI agreed to develop a special satellite for deployment. A science advisory team provided guidance on science accommodation requirements prior to the formal joint NASA-ASI Announcement of Opportunity for science investigations being issued in April, 1984.

The purpose of the TSS was to provide the capability of deploying a satellite on a long, gravity-gradient stabilized tether from the Space Shuttle where it would provide a research facility for investigations in space physics and plasma-electrodynamics. Nine investigations were selected for definition for the first mission (TSS-1) in July, 1985. In addition, ASI agreed to provide CORE equipment (common to most investigations) that consisted of two electron guns, current and voltage monitors and a pressure gauge mounted on the Orbiter, and a linear accelerometer and an ammeter on the satellite. NASA agreed to add a hand-held low light level TV camera, for night-time observation of the deployed satellite. The U.S. Air Force Phillips Laboratory agreed to provide a set of electrostatic charged particle analyzers, mounted in the Shuttle's payload bay, to determine Orbiter potential.

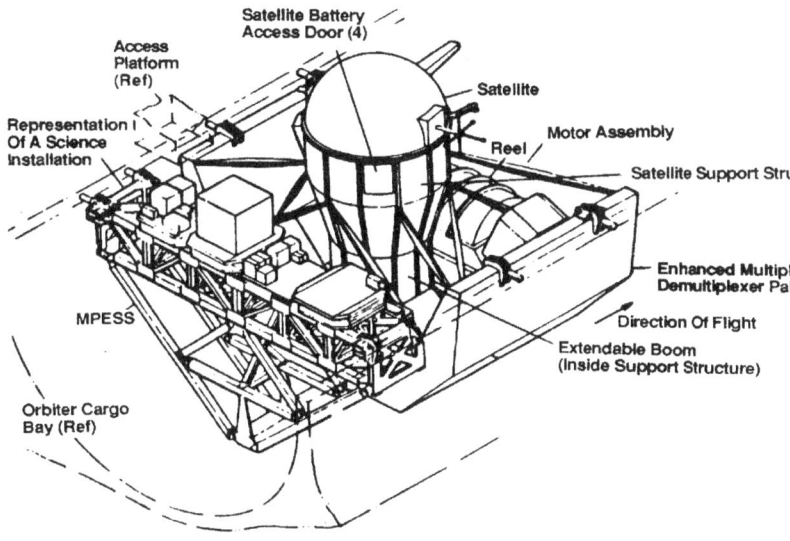

Figure 1.2   TSS-1 Configuration on Orbiter

During TSS-1, which was launched July 31, 1992 on STS-46, the Italian satellite was deployed 268 m directly above the Orbiter where it remained for most of the mission. This provided over 20 hours of stable deployment in the near vicinity of the Orbiter--the region of deployed operations that was of greatest concern prior to the mission. The TSS-1 results conclusively show that the basic concept of long gravity-gradient stabilized tethers is sound and settled several short deployment dynamics issues, reduced safety concerns, and clearly demonstrated the feasibility of deploying the satellite to long distances--which allowed the TSS-1R mission to be focused on science objectives.

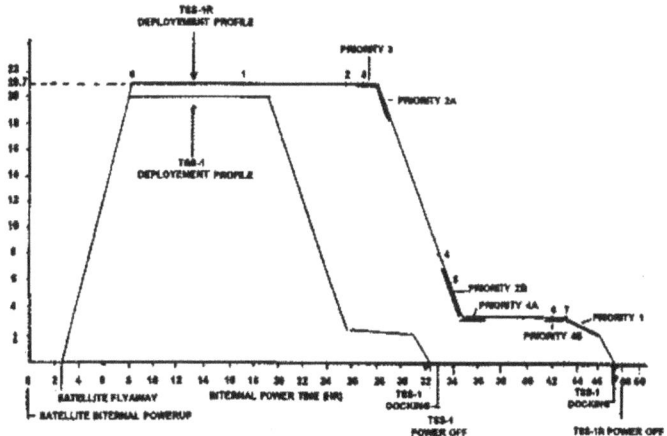

Figure 1.3   TSS1 and TSS1R Timelines

The TSS-1R mission was launched February 22, 1996 on STS-75. During this mission, the satellite was to have been deployed 20.7 km above the Space Shuttle on a conducting tether where it was to remain for more than 20 hours of science experiments, followed by a second stop for an additional seven to nine hours of experiments at a deployed distance of 2.5 km.

The goals of the TSS-1R mission were to demonstrate some of the unique applications of the TSS as a tool for research by conducting exploratory experiments in space plasma

3

physics. It was anticipated that the motion of a long conducting tether through the Earth's magnetic field would create a large motional emf that would bias the satellite to high voltages and drive a current through the tether system. The circuit for the tether current would be closed by a large external loop in the conducting ionospheric plasma where an array of physical phenomena and processes would be generated for controlled studies.

Although the TSS-1R mission was not completed as planned, the Italian satellite was deployed to a distance of 19.7 km--making TSS-1R the largest man-made electrodynamic structure ever placed in orbit. This deployment was sufficient to generate high voltages across the tether and extract large currents from the ionosphere. These voltages and currents, in turn, excited several space plasma phenomena and processes of interest. Active tether science operations had begun at satellite fly-away and continued throughout the deployment phase, which lasted more than 5 hours. As a result, a high-quality data set was gathered and significant science activities had already been accomplished prior to the time the tether broke. These activities included the measurement of the motional emf, satellite potential, Orbiter potential, current in the tether, charged particle distributions, and electric and magnetic fields. Significant findings include:

(1) Currents, collected by the satellite at different voltages during deployment, that exceeded the levels predicted by the best available numerical models by factors of up to three (see figure 1.4).

(2) Energetic electrons, that are not of natural ionospheric origin and whose energy ranged as high as 10 keV, were observed coincident with current flow in the tether. These data suggested possible energization of electrons by wave-particle interactions(see figure 1.5).

(3) A large increase of the tether current, a precipitous drop of the satellite bias voltage, very intense and energetic ion fluxes moving outward from the satellite's high-voltage plasma sheath, and a strong enhancement of the ac electric field in the 200 Hz to 2 kHz range-all observed to be concurrent with a satellite ACS yaw thruster firing. These observations imply a plasma density enhancement by ionization of the neutral gas emitted by the satellite thrusters.

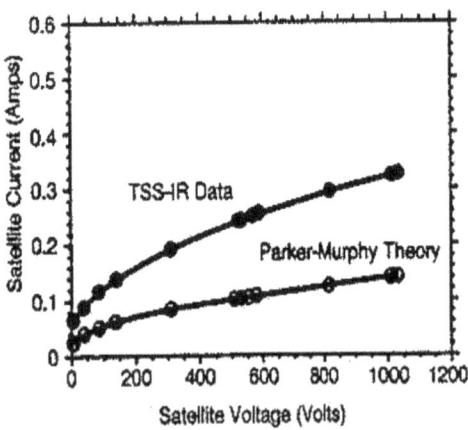

Figure 1.4 Measured TSS-1R and theoretically predicted I-V characteristics

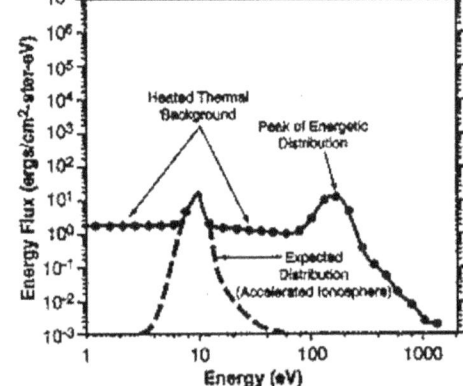

Figure 1.5 Energetic electron Population measured at the satellite's surface.

It is already apparent, therefore that the data gathered during TSS-1R have the potential to significantly refine the present understanding of the physics of (1) the collection of current and production of electrical power or electrodynamic thrust by high-voltage tethered systems in space, (2) the interaction of spacecraft, and even certain types of celestial bodies, with their local space plasmas, and (3) neutral gas releases in space plasmas and their effect on both of the above processes.

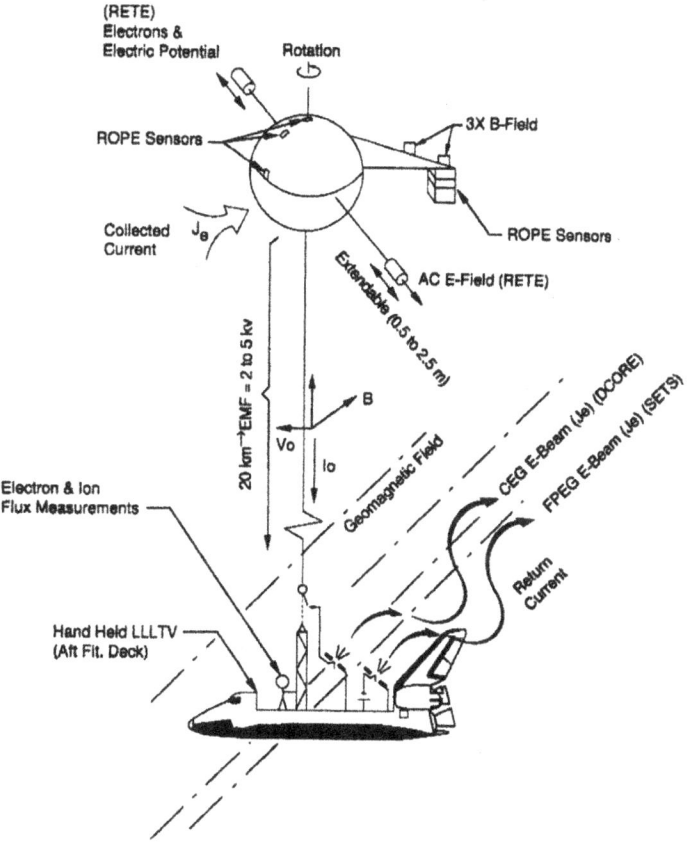

Figure 1.6 TSS Functional Schematic

The sensor package on the boom was electrically isolated from the satellite, and its potential was controlled by the ROPE floating power supply. For satellite potentials up to 500 V, the sensor package was maintained near the local plasma potential to allow unambiguous measurements to be obtained. The potential of the sensor package could also be swept to allow the package itself to serve as a diagnostic probe.

TSS-1R Science Investigators

TSS Deployer Core Equipment and Satellite Core Equipment (DCORE/SCORE)

Carlo Bonifazi, Principal Investigator
Agenzia Spaziale Italiana

The Tethered Satellite System Core Equipment will demonstrate the capability of a tethered system to produce electrical energy and will allow studies of the electrodynamic interaction of the tethered system with the ionosphere. The TSS Core Equipment controls the current flowing through the tether between the satellite and the orbiter and makes a number of basic electrical and physical measurements of the Tethered Satellite System.

Deployer Core Equipment consists of several instruments and sensors on the starboard side of the MPESS in the

5

. payload bay. A master switch connects the tether conductor to science equipment in the orbiter payload bay; a power distribution and electronic control unit provides basic power, command, and data interfaces for all Deployer Core Equipment except the master switch; and a voltmeter measures the tether potential with respect to the orbiter structure. The Core Electron Accelerator has two electron-beam emitters that can eject up to 750 milliamperes of current from the system. Two other instruments complement the electron accelerator's operations: a vacuum gauge to measure ambient gas pressure and to prevent operation if pressure conditions could cause arcing and a device to connect either generator head to the tether electrically.

Satellite Core Equipment consists of a linear three-axis accelerometer and an ammeter. The accelerometer (along with the satellite's gyroscope) will measure satellite dynamics, while the ammeter will provide a slow sampling monitor of the current collected on the skin of the TSS-1R satellite.

Research on Orbital Plasma Electrodynamics (ROPE)

Nobie Stone, Principal Investigator
NASA Marshall Space Flight Center

This investigation is designed to study the behavior of the ambient ionospheric charged particle populations and of ionized neutral particles around the TSS-1R satellite under a variety of conditions. Since the collection of free electrons from the surrounding plasma produces current in the tether, knowledge of the behavior of charged particles is essential to understanding the physics of tether current production.

From its location on the satellite's fixed boom, the Differential Ion Flux Probe measures the energy, temperature, density, and direction of ambient ions that flow around the satellite, as well as neutral particles that have been ionized in the satellite's plasma sheath and accelerated outward radially. In this instrument, an electrostatic deflection system, which determines the charged particle direction of motion over a range of 100 degrees, routes particles to a retarding potential analyzer, which determines the energy of the ion stream, measuring particle energies from 0 to 100 electron volts (eV). The directional discrimination of the Differential Ion Flux Probe will allow scientists to differentiate between the ionospheric ions flowing around the satellite and the ions that are created in the satellite's plasma sheath and accelerated outward by the sheath's electric field.

The Soft Particle Energy Spectrometer instrument is a collection of five electrostatic analyzers that measure electron and ion energies from 1 to 10,000 eV. Three analyzer modules provide measurements at different locations on the surface of the satellite's hemispherical Payload Module. These sensors determine the potential of the satellite and the distribution of charged particles flowing to its surface. Two other Soft Particle Energy Spectrometer sensors, mounted with the Differential Ion Flux Probe on the end of the boom, measure ions and electrons flowing both inward and outward from the satellite. These measurements can be used to calculate the local potential of the plasma sheath.

The sensor package on the boom is electrically isolated from the satellite, and its potential is controlled by the floating power supply. For satellite potentials up to 500 V, the sensor package will be maintained near the local plasma potential to allow unambiguous measurements to be obtained. The potential of the sensor package also can be swept, allowing the package itself to serve as a diagnostic probe.

Research on Electrodynamic Tether Effects (RETE)

Marino Dobrowolny, Principal Investigator
Agenzia Spaziale Italiana

The behavior of electrostatic waves and plasma in the region around a tethered satellite affects the ability of that satellite to collect ions or electrons and, consequently, the ability of the tether to conduct an electric current. This investigation provides a profile of the electrical potential in the plasma sheath and identifies waves excited by this potential in the region around the satellite. probes, placed directly into the plasma in the vicinity of the satellite, map alternating current (ac) and direct current (dc) electric and ac magnetic fields produced as the current in the tether is changed by instabilities in the plasma sheath or as the Fast-Pulse Electron Accelerator or Core Electron Accelerator or Core Electron Accelerator is fired in the payload bay.

The instruments are mounted in two canisters at the end of a pair of 2.4 m extendible booms. As the satellite spins, the booms are extended, and sensors measure electric and magnetic fields, particle density, and temperature at various angles and distances in the equatorial plane of the satellite. To produce a profile of the plasma sheath, measurements of dc potential and electron characteristics are made both while the boom is fully extended and as it is being extended or retracted. The same measurements, taken at only one distance from the spinning satellite, produce a map of the angular structure of the earth.

One boom carries a wave sensor canister, which contains a three-axis ac electric field meter and a two-axis search coil ac magnetometer to identify electric fields and electrostatic waves and to characterize the intensity of surrounding magnetic fields. Highly sensitive radio receivers and electric field preamplifiers within the canister complement the operations of the probes.

On the opposite boom, a plasma package determines electron density, plasma potential, and low-frequency fluctuations in electric fields around the satellite. A Langmuir probe with two metallic sensors samples the plasma current; from this measurement, plasma density, electron temperature, and plasma potential may be determined. This potential is then compared to that of the satellite. Two other probes measure low-frequency electric fields.

Magnetic Field Experiment for TSS Missions (TEMAG)

Franco Mariani, Principal Investigator
Second University of Rome

The primary goal of the TEMAG investigation is to map the magnetic fields around the satellite. If the magnetic disturbances produced by satellite interference, attitude changes, and the tether current can be removed from measurements of the ambient magnetic fields, the Tethered Satellite System will prove an appropriate tool for magnetic field studies.

Two triaxial fluxgate magnetometers, very accurate devices designed to measure magnetic field fluctuations, are located on the fixed boom. One sensor at the tip of the boom and another at mid-boom characterize ionospheric conditions at two distances from the satellite, determining the magnetic signature that is produced as the satellite moves rapidly through the ionosphere. Combining measurements from the two magnetometers allows real-time estimates to be made of the magnetic fields produced by the presence of satellite batteries, power systems, gyros, motors, relays, and permanent magnets. The environment at the tip of the boom should be less affected by the spacecraft subsystems than that at mid-boom. After the mission, the variable effects of switching satellite subsystems on and off, of thruster firings, and of other operations that introduce magnetic disturbances will be modeled by investigators in an attempt to remove these spurious signals from the data.

The two magnetometers will make magnetic field vector readings 16 times per second to obtain the geographic and temporal resolution needed to locate short-lived or thin magnetic structures. The readings will be made two times per second to allow discrimination between satellite-induced magnetic noise, the magnetic signals produced by the tether current, and the ambient environment. The magnetometers will alternate these rates: while the one on the tip of the boom operates 16 times per second, the midpoint magnetometer will operate twice per second and vice versa. Data gathering begins as soon as possible after the satellite is switched on in the payload bay and continues as long as possible during satellite retrieval.

Shuttle Electrodynamic Tether System (SETS)

Brian Gilchrist, Principal Investigator, University of Michigan

This investigation is designed to study the current-voltage characteristics of the orbiter-tether-satellite system and the fundamental controlling parameters in the Earth's ionosphere. This is accomplished through control of the tether system electrical load impedance and the emission of electrons at the orbiter end of the system. The experiment also explores the use of space tethers as science tools. Orbiter charging processes are measured using electron emissions plus the tethered satellite as a remote electrical reference. Plasma waves generated by electron beams are measured by receives at the satellite. Ionospheric spatial structure is investigated by simultaneous in-situ measurements at both the orbiter and satellite. Also, electrodynamic tether low-frequency radio wave reception, emission, and transient response are investigated.

The hardware is located on the MPESS near the center of the payload bay and adjacent to the deployer pallet. A Tether Current-Voltage Monitor measures tether current and voltage, while controlling tether circuit load resistance. The Fast-Pulse Electron Accelerator emits an electron beam of 100 or 200 milliamperes at an energy of 1000 electron volts. The beam can be pulsed with on/off times ranging from 400 nanoseconds to 107 seconds. The beam balances the tether current and is used to control the level of charging of the Space Shuttle orbiter. In addition, the beam is used as an active stimulus of the plasma near the orbiter in support of several scientific objectives.

The Spherical Retarding Potential Analyzer, mounted on a stem at one corner of the support structure, records plasma ion density and energy distribution in the payload bay. Similarly, a Langmuir Probe measures electron plasma temperature and density and is mounted on the tower also. At the center of the support structure, the Charge and Current Probe measures the return current to the orbiter, recording large and rapid changes in orbiter potential, such as those that are produced when electrons are conducted from the tether to the orbiter frame or when an electron beam is emitted. A three-axis fluxgate magnetometer measures the magnetic field, allowing the magnetic field lines in the payload bay to be mapped, which is crucial since electron beams and the flow of plasma spiral in response to these fields. Using this information, scientists can aim the electron beam at various targets, including orbiter surfaces, to study the fluorescing that occurs.

Shuttle Potential and Return Electron Experiment (SPREE)

David Hardy, Principal Investigator
Department of the Air Force, Phillips Laboratory

SPREE will measure the charged particle populations around the orbiter for ambient space conditions and during active TSS-1R operations. SPREE supports the TSS-1R electrodynamic mission by determining the level of orbiter charging with respect to the ambient space plasma, by characterizing the particles returning to the orbiter as a result of TSS-1R electron beam operation, and by investigating local wave particle interactions produced by TSS-1R operations.

SPREE is mounted on the port side of the MPESS. The sensors for SPREE are two pairs of electrostatic analyzers, each pair mounted on a rotary table motor drive. The sensors measure the flux of all electrons and ions in the energy range from 10 eV to 10 keV that impact the orbiter at the SPREE location. The energy range is sampled either once or eight times per second. The sensors measure the electrons and ions simultaneously over an angular field of view of 100 x 10 degrees. This field of view, combined with the motion of the rotary tables, allows SPREE measurements over all angles out of the payload bay.

The Data Processing Unit (DPU) performs all SPREE command and control functions and handles all data and power interfaces to the orbiter. In addition, the DPU processes SPREE data for use by the crew and the ground support team. A portion of the SPREE data is downlinked in real time, and the full data set is stored on two SPREE Flight Data Recorders (FDRs). Each FDR holds up to 2 gigabytes of data for postflight analysis.

Tether Optical Phenomena Experiment (TOP)

Stephen Mende, Associate Investigator
Lockheed

Using a hand-held camera system with image intensifiers and special filters, the TOP investigation will provide visual data that may allow scientists to answer a variety of questions concerning tether dynamics and optical effects generated by TSS-1R. In particular, this experiment will examine the high-voltage plasma sheath surrounding the satellite.

In pace of the image-intensified conventional photographic experiment package that has flown on nine previous Shuttle missions, a charge-coupled device electronic system will be used instead of film. This new system combines the image intensifier and the charge-coupled device in the same package. The advantage of charge-coupled devices over film is that they allow real-time observation of the image, unlike film, which has to be processed after the mission. The system also provides higher resolution in low-light situations than do conventional video cameras.

The imaging system will operate in four configurations: filtered, interferometric, spectrographic, and filtered with telephoto lens. The basic system consists of a 55 mm F/1.2 or 135 mm F/2.0 lens attached to the charge-coupled device equipment. Various slide-mounted filters, an air-spaced Fabry Perot interferometer, and spectrographic equipment will be attached to the equipment so that the crew can perform various observations.

In one mode of operation, the current developed in the Tethered Satellite System is closed by using electron accelerators to return electrons to the plasma surrounding the orbiter. The interaction between these electron beams and the plasma is not well understood. Scientists expect to gain a better understanding of this process and how it affects both the spacecraft and the plasma by using the charge-coupled device to make visual, spectrographic, and interferometer measurements. Thruster gasses also may play a critical role in Tethered Satellite System operations. By observing optical emissions during the buildup of the system-induced electromotive force (emf) and during gas discharges, scientists can understand better the interaction between a charged spacecraft and the plasma environment and will increase their knowledge of how the current system closes at the poles of the voltage source.

Investigation of Electromagnetic Emissions by the Electrodynamic Tether (EMET)

Robert Estes, Principal Investigator
Smithsonian Astrophysical Observatory (SAO)

Observations at the Earth's Surface of Electromagnetic Emissions by TSS (OESEE)

Giorgio Tacconi, Principal Investigator
University of Genoa

One goal of these investigations is to determine the extent to which waves that are generated by the tether interact with trapped particles and precipitate them. Wave-particle interactions are thought to occur in the Van Allen radiation belts where waves, transmitted from Earth, "jar" regions of energetic plasma and cause particles to "rain" into the lower atmosphere. Although poorly understood, wave-induced precipitation is important because it may affect activity in the atmosphere closer to Earth. Various wave phenomena that need to be evaluated are discrete emissions, lightning-generated whistlers, and sustained waves, such as plasma "hiss." Wave receivers on the satellite detect and measure the characteristics of the waves, and particle detectors sense wave-particle interactions, including those that resemble natural interactions in radiation belts. Ground stations may be able to detect faint emissions produced as waves disturb particles and enhance ionization. Furthermore, the current is carried away from the tethered system through the ionosphere by electromagnetic waves. Also, investigators want to know what type of wave predominates in this process and whether the tether-ionosphere current closure occurs near the system or hundreds of kilometers away. Ground-based measurements may be able to shed light on this question.

Another goal is to determine how well the Tethered Satellite System can broadcast from space. Ground-based transmissions, especially those below 15 kHz, suffer from inefficiency. Since large portions of ground-based antennas are buried, most of the power supplied to the antenna is absorbed by the ground. Because of the large antenna size and consequent high cost, very few ground-based transmitters operate at frequencies below 10 kHz. Since the Tethered Satellite System operates in the ionosphere, it should radiate waves more efficiently. For frequencies lower than 15 kHz, the radiated signals from a 1 kW space transmitter may equal that from a 100 kW ground transmitter.

Waves generated by the tether will move in a complex pattern within the ionosphere and into the magnetosphere. EMET and OESE science teams will operate ground stations equipped with magnetometers at remote sites along the TSS ground track. The EMET sites on Mona Island (Puerto Rico) and Bribie Island (Australia) are capable of measuring frequencies from near dc to 40 kHz. The OESEE sites in the Canary islands and Kenya utilize Superconducting Quantum Interference Devices (SQUIDs) and coil magnetometers to monitor frequencies below 100 Hz. Researchers at these sites will try to measure the emissions produced by the TSS and will track the direction of waves that are generated when electron accelerators in the orbiter payload pulse the tether current as the orbiter passes overhead. The incoherent scattering radar and antenna at the Arecibo Radio Telescope facility will attempt to observe the ionospheric perturbations produced by the TSS system.

Investigation and Measurement of Dynamic Noise in the TSS (IMDN)

Gordon Gullahorn, Principal Investigator
Smithsonian Astrophysical Observatory

Theoretical and Experimental Investigation of TSS Dynamics (TEID)

Silvio Bergamaschi, Principal Investigator
Institute of Applied Mechanics

TSS-1R will be the longest structure ever flown in space, and its dynamic behavior will involve oscillations over a wide range of frequencies. Although the major dynamic characteristics are readily predicted, future applications of long tethers demand verification of the theoretical models. Moreover, higher frequency oscillations, which are essentially random, are more difficult to predict. This behavior, called "dynamic noise," is analogous to radio static. An understanding of its nature is needed if tethered platforms are to be used for microgravity facilities and for studying fluctuations in the small-scale structure of Earth's gravitational and magnetic fields. These gravitational fluctuations are caused by variations in the composition and structure of Earth's crust and may be related to mineral sources.

These two investigations will analyze data from a variety of instruments to study Tethered Satellite System dynamics. The primary instruments will be the accelerometers and gyros on board the satellite; however, tether tension and length measurements and magnetic field measurements also will be used. The dynamics will be observed in real time at the Marshall Space Flight Center (MSFC) Payload Operations Control Center (POCC) and will be subjected to detailed postflight analysis. Basic models and simulations will be verified (and extended or corrected as needed); then, these can be used confidently in the design of future tethered missions, both of the Tethered Satellite System and of other designs. The dynamic noise inherent to the system will be analyzed to determine if tethered systems are suitable for sensitive observations of the geomagnetic and gravitational fields and, if required, to develop possible damping methods.

Theory and Modelling in Support of Tethered Satellite Applications (TMST)

Adam Drobot, Principal Investigator
Science Applications International Corporation (SAIC)

This investigation will develop numerical models of the tether system's overall current and voltage characteristics, of the plasma sheaths that surround the satellite and the orbiter, and of the system's response to the operation of the electron accelerators. Also of interest are the plasma waves generated as the tether current is modulated. All data collected on the mission will be combined to refine these models.

Two- and three-dimensional mathematical models of the electrodynamics of the tether system will be developed to provide an understanding of the behavior of the electric and magnetic fields and the charged particles surrounding the satellite. These studies are expected to model the plasma sheath (through which the satellite travels) under a variety of conditions. This includes those in which the motion of the tether and neutral gas emissions from the thrusters are not considered, those that incorporate the effects of tether motion, and those that factor in the gas emissions.

The sheath surrounding the orbiter has several unique features that are related to the ability of the electron accelerators to control the orbiter's potential. Models of the orbiter's sheath, when small currents are flowing in the tether, will consider the potential of the orbiter to be negative; for large currents, models will be developed assuming a positive orbiter potential. In this way, the sheath structures and impedance characteristics of the orbiter/plasma interface can be studied.

The response of plasma to the electromotive force produced by the motion of the tether system through the geomagnetic field is another focus of the TMST investigation. Using data from other studies, kinetic plasma processes will be analyzed or numerically simulated by computer to model the reaction of the ionosphere to the passage of TSS-1R.

This investigation also models the relationship between the efficiency of wave generation and the amount of current flowing through the tether to examine how the tether antenna couples to the ionosphere and how ultra-low-frequency (ULF) and very-low frequency (VLF) wave propagate through the ionosphere. These models will complement the information gathered by TSS-1R instruments at ground stations.

The Subsatellite element (TSS-S) of the Tethered Satellite System (TSS)

Figure 1.7   TSS-S during TSS-1R Mission

The TSS-S is a Shuttle-tethered instrumented platform supporting dynamic and electrodynamic investigations; it thus avails the unique opportunity offered by the tether complex. The TSS-S has flown twice, first in August, 1992, then in February, 1996 and its performance has exceeded expectations both times.

As shown in figure 1.7, the Satellite has a roughly spherical shape with an outer diameter of 1.6 m; it features two fixed and two deployable/retractible booms. One of the fixed booms (with struts) is one meter long and it is meant for scientific instrument accommodation at its tip and midpoint (2.5 kg overall), while the other fixed boom supports the subsatellite's RF communications antenna. The two deployable/retractible booms (DRBs) are designed to take science instrument packages weighing up to 1.5 kg per boom up to 2.35 m away from the satellite shell in 14 mm steps.

As shown in fig. 1.8, the satellite is functionally divided into three "modules", namely the Service Module (SM), the Auxiliary Propulsion Module (APM) and the Payload Module (PM). The SM is the hemisphere located on the tether' side and it accommodates all subsystems but for the power and command-data handling units interfacing with the experiments; these, together with the science experiment equipment, are housed inside the PM, i.e. the hemisphere opposite to the tether. The SM and PM are separated by, and join at, the APM, which is made up by the equatorial plane, the propellant tank and all the valves, piping and propellant management equipment. The TSS Satellite has an overall mass of about 521 kg, out of which up to 66 kg made up by science instruments and 61 kg by the gaseous nitrogen propellant (GN2) for satellite attitude and rate control and for tether tension augmentation to support early TSS-S deployment and to keep the tether taut during proximity operations, when the gravity-gradient-originated tension is too weak to guarantee that the tether does not become slack. Yaw thrusters are provided at the Satellite's equatorial plane for yaw attitude and yaw rate control; each yaw thruster has two nozzles and provides 0.5 Nm pure torque about the "vertical" axis using about 1.7 g/s of on-board propellant. Yaw attitude control accuracy is about 1 deg of the desired angle while yaw rate control is accurate to +/- 0.1 RPM for rates in the range - 0.7 to + 0.7 RPM. The reference yaw angle and rate as well as the associated control deadbands can be selected by telecommand.

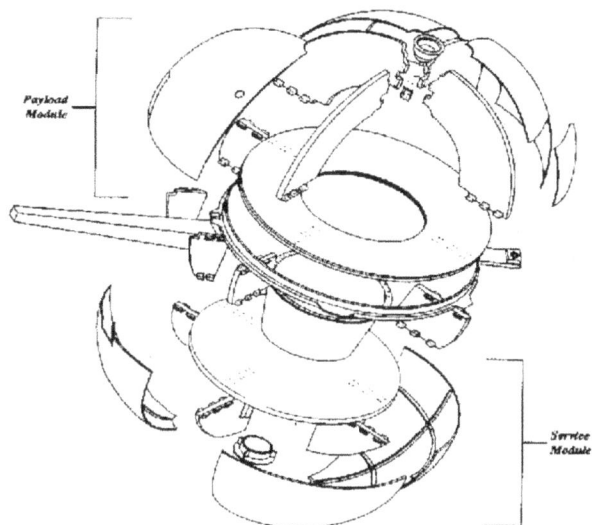

Figure 1.8   TSS-S Exploded view

Thrusters are also present close to the equatorial plane to control pitch and roll oscillation rates; in-plane and out-of-plane thrusters control the pitch and roll rates, respectively; they provide a 0.8 Nm torque about the relevant axis. The in-plane (pitch control) thrusters also generate pure forces along the x (roll) and z (yaw) axes, about 0.7 and 1 N in magnitude, respectively. Likewise, the out-of-plane (pitch control) thrusters give rise to pure force components along the y (pitch) and z axes, about 1.9 and 1 N in magnitude, respectively. In-plane and out-of-plane thrusters use about 4.4 and 3.8 g/s of GN2 each, respectively, and can be actuated one at a time only. They can operate under external command or under control from the TSS-S on board software in the so-called Auto Rate Damping (ARD) mode. The TSS-S's tether-aligned thrusters, in-line 1 and 2, each providing 2 N pure thrust along the z axis, use about 3.2 g/s when active and can be actuated upon external command either individually or together. They are meant for tether tension augmentation and support TSS-S separation from the Orbiter during early deployment and close-in approach to the Orbiter during final retrieval.

The Satellite is provided with a complete set of attitude detrmination sensors, i.e. 4 rate-integrating gyroscopes, two bolometer-based optical Earth sensors (ES) and four Digital Sun Sensors (FDSS). Satellite attitude determination is carried out on board he satellite with a +/- 1 deg accuracy whenever the satellite is in attitude hold mode; the on-board attitude determination algorythm is based on gyroscope output and makes use of ES output for gyro drift compensation. Ground-based algorythms have been developed by Alenia Spazio to more accurately reconstruct the Satellite attitude history, even while in spin and passive mode, to support post-flight science data analyses; under normal operating conditions and data availability, they can provide TSS-S attitude history reconstruction to better than 1'.

The TSS Satellite element also carries on board a set of four Ag-Zn batteries to support the deployed mission; they can provide up to 10.6 kWh, depending on their discharge profile and thermal conditions, as ascertained by both ground testing and flight experience. Out of the total, science experiments are allocated about 2.5 kWh overall, with a 100 W maximum overall power level. Twelve individually switched and fused power lines are provided for use

13

by the TSS-S science experiments, 4 with 5 A rating and 8 with 1.5 A rating, at 30 +/- 6 V input voltage.

The TSS-S provides a 16 kbps continuous telemetry stream, out of which about 4 kbps subsystem housekeeping, 10.25 kbps science instrument telemetry (housekeeping and science data) and about 1.75 kbps service (sunchronisation) words. Discrete, analog and 16-bit serial monitors can be acquired from the science instruments and inserted into the telemetry stream; analog monitors are A/D converted to 8-bit words. The TSS-S supports a 2 kbps maximum telecommand bitrate; the corresponding telecommand rate depends on whether the telecommands require processing by the Satellite on-board software, and can reach the maximum value of about 20 commands/s in case no processing is required. Science experiments can be provided relay-driving, discrete commands as well as 16-bit serial digital commands; no OBDH processing is provided on science experiments commands but routing to the end user. The TSS-S has a 40-slot Time Tagged Command Buffer (TTB), where commands can be stored for execution at a later time. Out of these, up to 30 can be allocated to the science experiments; TTB time tag resolution is about 32 sec, i.e. time tags can differ by 32 sec as a minimum, but commands with the same time tag are executed within 128 msec of each other in a FIFO order.

Besides engineering resources and capabilities, the experiments on board the TSS-S are provided with a magnetic cleanliness program which ensures that DC magnetic field generated by the satellite does not exceed about 20 nT at the fixed boom tip, with a very high stability (a few nT). Additionally, the TSS-S outer shell is coated with a 100-120 micron-thick conductive paint layer applied directly on the shell bare metal (Al); the paint helps keeping the resistance opposed by the skin to the electric current flow to a few tens of Ohms, the exact value depending on paint thickness and applied voltage. Ground testing and flight data have proved both the magnetic cleanliness level and the TSS-S overall conductivity to match or exceed the science requirements.

The TSS-S is equipped with two "standard" science support equipment items, namely the Satellite Ammeter (SA) and the Satellite Linear Accelerometer (SLA). The SA is a four scale (± 5, ± 0.5, ± 0.1, ± 0.02 A), auto-ranging instrument capable of providing measurements of the electric current flowing in the tether with a 7-bit accuracy over each range; SA data are provided 16 times a second in the satellite telemetry stream. The instrument, however, also has a 1 kHz bandwidth analog output, allowing other satellite experiments to directly sample current impulse waveforms. The SLA is a three-axis accelerometer with inductive-mechanical (coil-spring) control loop and capacitive pick-off; the instrument provides three mutually orthogonal acceleration measurements in the range -60 - +20 milli-g (z axis) and -20 - +20 milli-g (x, y axes), each available 16 times a second inside the satellite telemetry, with accuracies ranging from 100 micro-g (z axis) to 10 micro-g (x, y axes). The instrument measurement bandwidth is 4.5 Hz.

The experience acquired with the two performed flights has allowed very accurate characterisation of all TSS-S performance characteristics and has provided Alenia Spazio with expertise and S/W tools which allow the Company to provide in-depth and extensive support to both dynamics and electrodynamics analyses as well as to mission analysis, preparation and support.

Contacts for the TSS Project:

∑ M. Calabrese, R. Carovillano, T. Stuart - NASA Hqts.
∑ C. Bonifazi, M.Dobrowolny - ASI
∑ F.Giani, B.Strim, - Alenia
∑ N.Stone, R. McBrayer - NASA/MSFC
∑ TSS Investigator Working Group

1.2 The Small Expendable Deployer System (SEDS): SEDS-1 and SEDS-2 Missions

The SEDS project started as a Small Business Innovative Research contract awarded to Joe Carroll by NASA MSFC. SEDS hardware proved to be able to succesfully deploy a 20 km tether in space. Both flights of SEDS-1 (March 29, 1993) and SEDS-2 (March 9, 1994) flew as secondary payloads on Delta II launches of GPS satellites. After the third stage separation the end-mass was deployed from the second stage. SEDS-1 demonstrated the capability of deorbiting a 25 kg payload from LEO. SEDS-2, on the other end, demonstrated the use of a closed loop control law to deploy a tethered payload along the local vertical.

SEDS' hardware, as shown in figure 1.9, consists of a deployer, brake/cutter and electronics box. All the components that are in contact with the tether, except for the brake post, are coated with teflon. The deployer consists of baseplate, core, tether and canister. The tether is wound around the core. In addition there are three Light Emitting Diodes (LED). Two of the LED's are used to count the turns of deployed tether, while the third is used to check when the tether is almost completely unwound. The canister provides a protective cover for the tether and restrains it during deployment. The tether material is SPECTRA-1000.

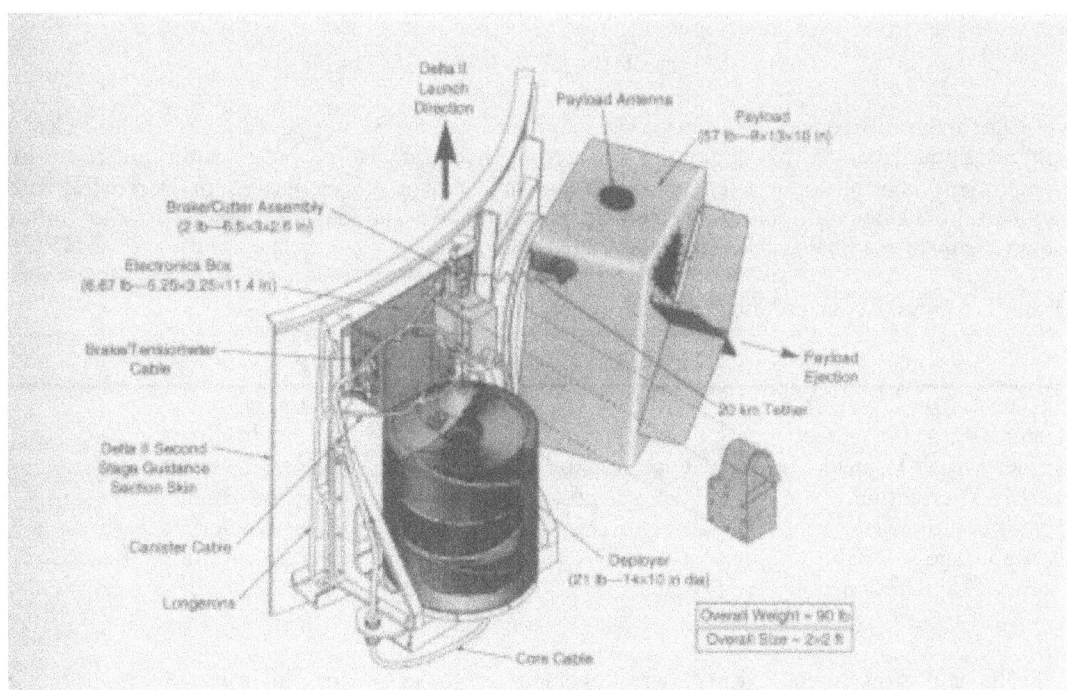

Figure 1.9 SEDS and Endmass on the Delta Second Stage

The brake/cutter components are: brake post, stepper motor, tensiometer, temperature sensor, pyro cutter, exit guide. The tether post is coated with hard anodize. The stepper motor is used to wrap or unwrap the tether to vary the deployment tension and the resulting deployment velocity. The brake mechanism is a friction multiplier and the multiplier function is proportional to the friction surface area between the tether and brake post. SEDS functional diagram is shown in figure 1.10.

15

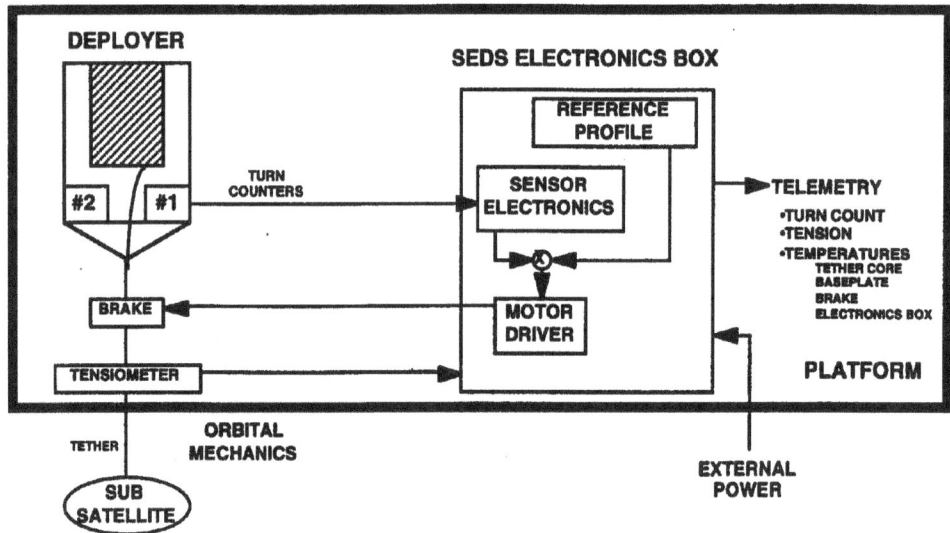

Figure 1.10 SEDS Functional Diagram

The main differences between SEDS 1 and SEDS-2 are shown in table 1. SEDS-2 closed loop was implemented by deploying the tether according to a pre-mission profile. The deployment control logic acted on the brake mechanism by increasing or decreasing the deployment velocity to follow the profile and bring the payload at the end of the tether deployment to a smooth stop along the local vertical.

Table 1. Main differences between SEDS-1 and SEDS-2

	SEDS-1	SEDS-2
Tether Cutter Pyrotechnics	Active	Inactive
Control Law	Open Loop	Closed Loop
Tether Solder Lumps	Study Tension Pulses	None
Tether Fabrication	Tether Application	Cortland/Hughes
Mission Initiation	Prior to Depletion Burn	After Depletion Burn
Brake Usage	Minor	Significant after 1 Km
Tether Stabilization	None	Yes

The end-mass payload (EMP) was developed by NASA LaRC in order to monitor the dynamics of a tethered susbsatellite. EMP consisted of three primary science sensors: a three-axis accelerometer, a three axis tensiometer and a three axis magnetometer. The EMP measured 40.6X30.5X20.3 cm and weighted about 26 kg. The end-mass was completely autonomous and carried its own battery, electronics, computer and S-band telemetry system. As schematic of EMP is shown in fig. 1.11. The three axis tensiometer was also developed at NASA LaRC.

SEDS-1 mission objectives were to demonstrate that SEDS hardware could be used to deploy a payload at the end of a 20 km-long tether and study its reentry after the tether was cut. The orbit chosen had an inclination of 34 degrees and a perigee altitude of 190 km and

16

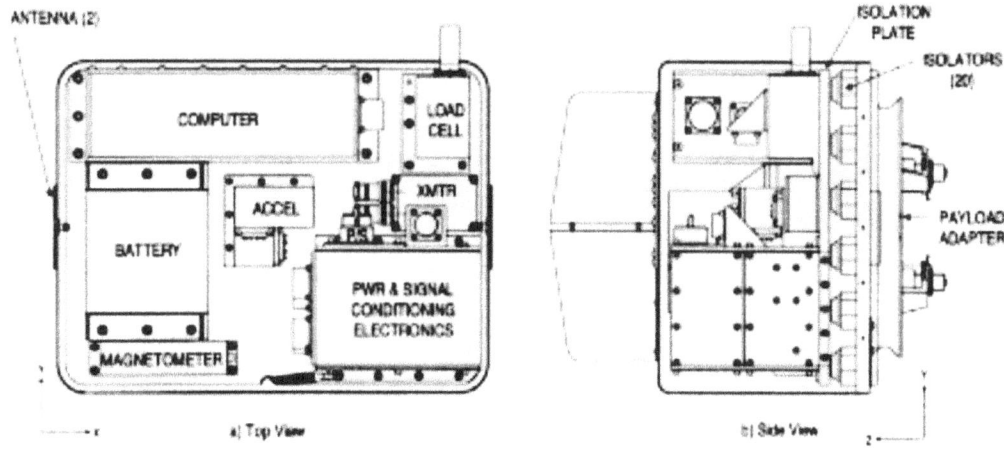

Figure 1.11 Schematic of SEDS EMP

an apogee altitude of 720 km. The EMP transmitted over 7900 seconds of data before burning into the atmosphere (1Hz sampling rate for the magnetometer and 8 Hz for the tensiometers and accelerometers). As predicted, SEDS-1 reentry was off the coast of Mexico (see fig. 1.12a). NASA stationed personnel at Cabo San Lucas, Puerto Vallarta and Manzanillo to make photographic and video observations. The Puerto Vallarta site was able to obtain observational data as shown in figure 1.12b

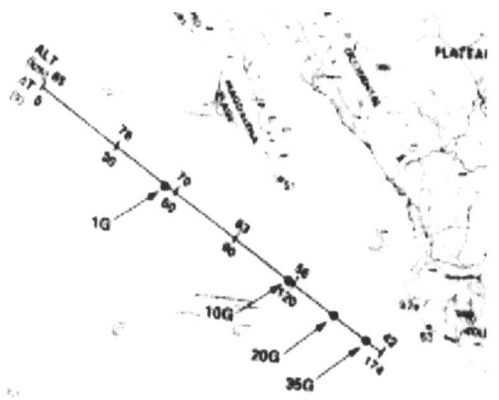

Figure 1.12a SEDS-1 EMP reentry trajectory

Figure 1.12b Observational Data of SEDS-1 reentry

SEDS-2 mission objectives were to demonstrate the feasibility of deploying a payload with a closed-loop control law (i.e. a predetermined trajectory) and bring it to a small final angle (<10 degrees) along the local vertical. A secondary objective was to study the long term evolution of a tethered system. The orbit this time was chosen to be circular with an altitude of about 350 km. The SEDS-2 tether was allegedly cut by a micrometeroid or debris after five days. The EMP transmitted over 39,000 seconds of data before the battery died (1 Hz sampling rate for all the three primary science sensors).

SEDS-1 and SEDS-2 Flight Data

SEDS data base is available through anonymous ftp at the node optimu@gsfc.nasa.gov (128.183.76.209) SEDS1 data are in the subdirectory /pub/projects/tether/SEDSMission1 and SEDS-2 data are in the directory /pub/projects/tether/SEDSMission2. Each directory is organized in different subdirectories with deployer data, EMP data, radar, etc.. Each content of a directory is described in a read.me file.

SEDS-1

The turn counter data are shown in Figure 1.13a, the tension at the deployer is shown in figure 1.13b and the tether rate in 1.13c. In order to compute the tether length and its rate, the turns had to be mapped and converted into deployed length. Note that the velocity at the end of the deployment was about 7 m/s explaining the huge jump in tension and the consequent rebounds.

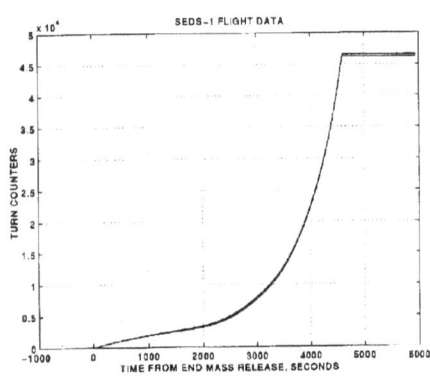

Figure 1.13a. SEDS Deployer Turns Counts

Figure 1.13b. SEDS Deployer Tension

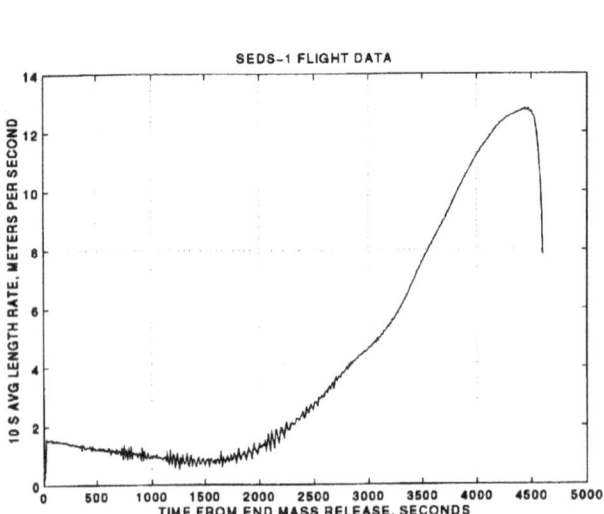

Figure 1.13c. SEDS Deployer 10-sec Average Length Rate

18

The magnetometer and tension moduli at the EMP are shown in figures 1.13d and 1.13e, respectively. Note that the magnetometer was affected by a bias estimated to be 3065 nT, -3355 nT and -4188 nT on the x, y and z axes, respectively. Procedures on the data calibration and validation are given at the ftp site as well as are described in several papers presented at the Washington Conference.

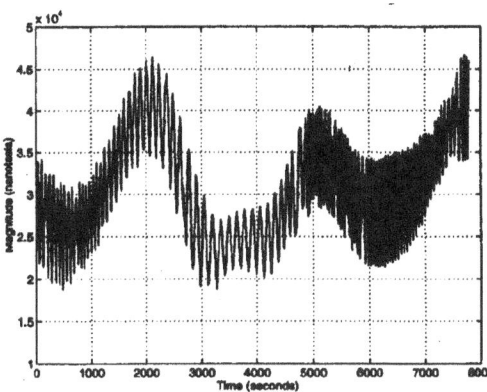

Figure 1.13d. EMP Magnetometer Modulus

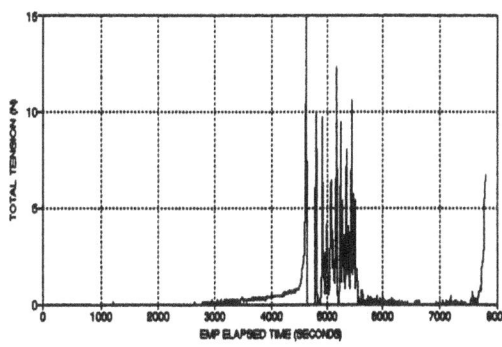

Figure 1.13e. EMP Tension Modulus

SEDS-2

The tether deployment rate and the tension at the deployer are shown in figure 1.14a, and 1.14b, respectively. The deployment law was so effective that the final tether rate was about 2 cm/s. As computed by the modulus of the EMP tension, shown in fig 1.14c, the final libration was about 4 degrees, and it was confirmed also by the radar tracking. Even in SEDS-2 the magnetometer signal was affected by a bias anomaly that was estimated to be -1128 nT, 1312 nT, and 2644 nT on the x,y and z axes, respectively.

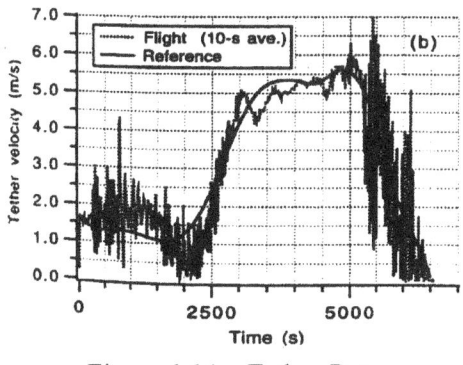

Figure 1.14a. Tether Rate

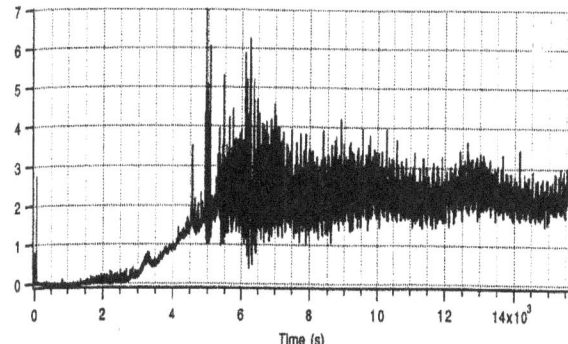

Figure 1.14b. Tension at Deployer

19

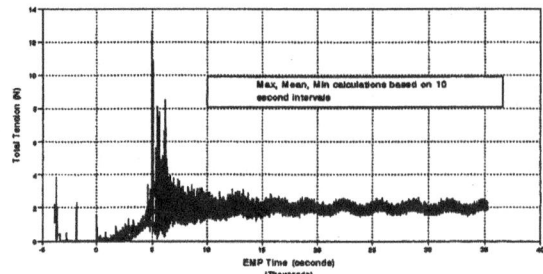

Figure 1.14c. EMP Tension Modulus

Contacts for the SEDS Project:

- J.Harrison , H.Frayne Smith, K.Mowery, C.C. Rupp - NASA/MSFC
- J.Carroll - Tether Applications
- J. Glaese - Control Dynamics
- M.L. Cosmo, E.C.Lorenzini, G.E. Gullahorn -SAO
- T.Finley ,R.Rhew, J.Stadler - NASA/LaRC
- W.Webster - NASA/GSFC

1.3 The Plasma Motor Generator (PMG)

The PMG experiment was designed to test the ability of a hollow cathode assembly (HCA) to provide a low impedance bipolar electrical current between a spacecraft and the ionosphere. The 500m-long tether was chosen to assure complete separation between the grounded ends, forcing current closure through the ionosphere rather than with local overlap of the two plasma clouds. In order to function properly, an electrodynamic tether needs to be effectively "grounded" on both ends. The experiment aimed at demonstrating that such configuration could function either as a orbit-boosting motor or as a generator converting orbital energy into electricity, as shown in figure 1.15.

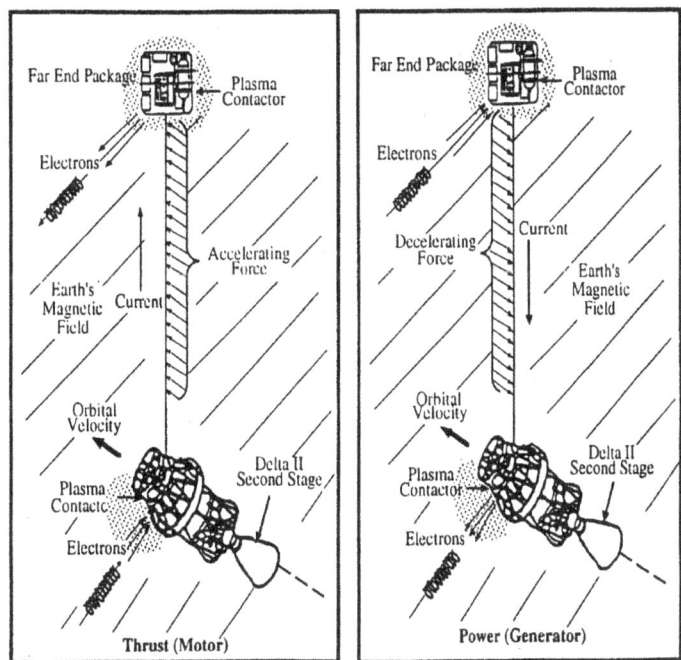

Figure 1.15 PMG investigation of an electrodynamic tether

The mission objectives were:

- HCA Operation
- End-mass separation greater than 200 m
- Induced voltage of 30 V or higher
- PMG Plasma clouds completely separated
- Achieve currents in the 0.1 -1 Amp
- Reverese current in tether using bias voltage
- Observe tether stability for gravity gradient vs. $\underline{I}X\underline{B}$ forces
- Collect I-V Characteristics for full orbit

As shown in figure 1.16, PMG consisted of four major subsystems: The Far-End Package (FEP), the Near-End Package (NEP), an electronics box (SEDS) and the Plasma Diagnostic Package (PDP). The system was launched as a secondary payload on a Delta II on June 26,

21

1993. After the third stage separation, PMG was left in an elliptical orbit (193X869) at 25.7 deg inclination. The FEP was ejected upward with an initial velocity of about 2-3 m/s. PMG was programmed to operate in three different data modes, by using a microprocessor to control selectable load resistors, to change 1) bias voltage levels, 2) polarity reversal 3) bypass relays.

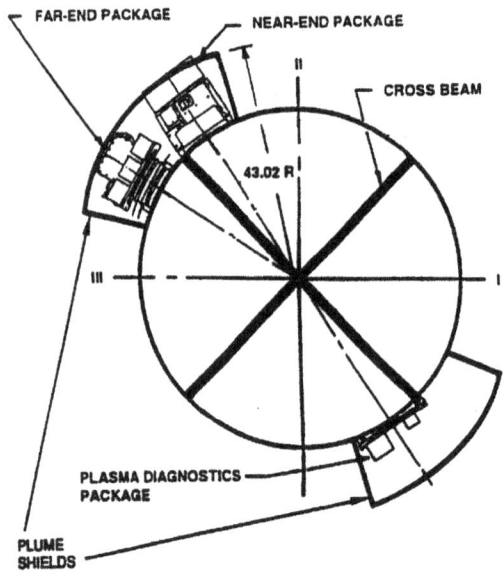

Figure 1.16 PMG Major hardware components

The SEDS deployer fixed spool concept was adapted for use, without the brake mechanism to provide minimum friction deploymnet of the relatively massive tether. The PDP experiment, developed by NASA/LeRC/U. of New Hampshire, was added in order to measure the deployer potential.

The NEP included a power "ON" relay, a microprocessor based control/data module and electrometer, and a tether bias voltage power supply.

Each HCA was equipped with a 1 liter gas bottle, on/off solenoid, gas metering block and power supplies to produce a weakly ionized xenon cloud. Both end platforms carried a 28 volt silver cell battery for a nominal 3-6 hours lifetime. The tether was a #18 AWG teflon insulated copper wire.

After deployment, during the first 150 minutes, sets of I vs. V performance data were obtained by applying bias voltages of +65V to -130V in series with the **IXB** induced emf, while varying load resistance in steps from 200 to 700 ohms total tether current path internal resistance (see figure 1.17). Total tether voltage was measure by placing a 2.2 MOhms resistance in series with the tether.

The PMG current showed to be fully reversible, operating either as a generator system with electron current flow down the tether or as a motor with electron current driven up the tether.

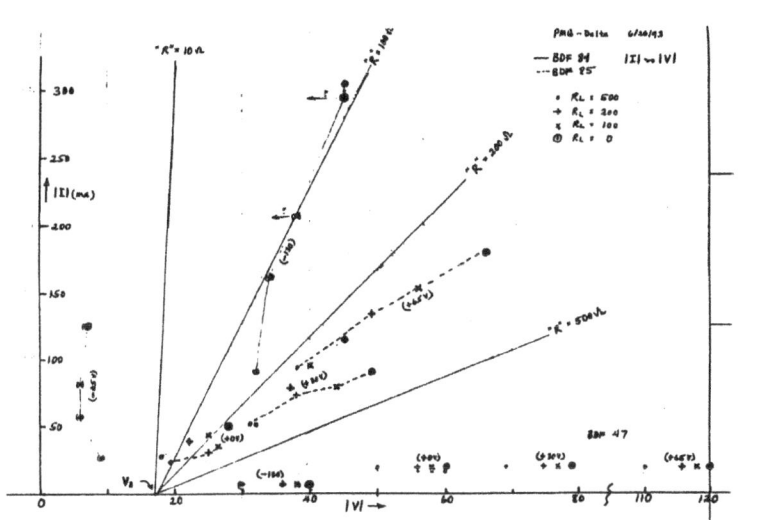

Figure 1.17 PMG I-V Curves

The data confirmed that the HCAs were able to complete PMG's current loop and 100-300 mA currents were observed in the daytime portion (probably due to the enhanced plasma denity) of the orbit and 10-50 mA on the nightime side, as shown in figure 1.18.

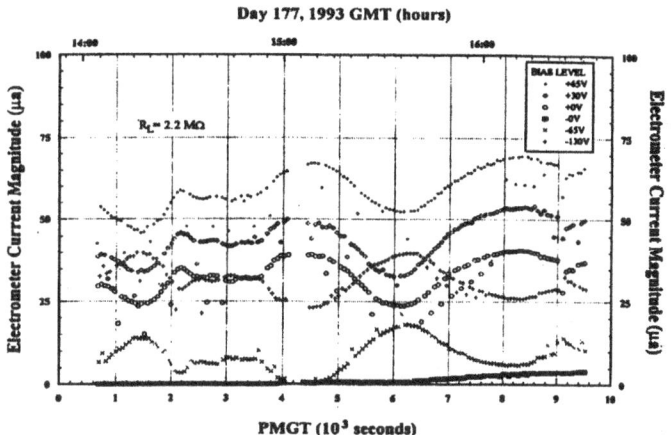

Figure 1.18 PMG Electrometer Reading of 2.2 MOhms Load and all bias voltages

The induced emf was measured with total voltage biased from +150 to -90 volts, and also with the bias turned off, by using only the induced emf. Variability of the induced emf has been matched against different models of electrodynamic interactions.

A contingent study was the detection with ground-based radars and squid magnetometers (see OESEE experiment on TSS1, G. Tacconi PI) of the plasma disturbances, ELF waves radiated by the system, the HCA plasma clouds and their associated plasma/ionosphere currents.

The experiment duration, in terms of plasma contactor operation and consequential active environment interaction, lasted about seven hours, until the batteries expired.

PMG data base is available through anonymous ftp at the node optimu@gsfc.nasa.gov (128.183.76.209) under the directory /pub/projects/tether/PlasmaMotorGenExp.

Contacts for the PMG Project:

- J.McCoy - NASA/JSC
- J.Carroll - Tether Applications
- M.D.Grossi, R.Estes -SAO
- R.J.Jost - System Planning Corporation
- R.C. Olsen -Naval Postgraduate School
- I.Katz - S-Cubed
- G. Tacconi, L. Minna - U. of Genoa
- D.C. Ferguson -NASA/LeRC
- R.Tolbert -U. of New Hampshire
- W.Webster - NASA/GSFC

1.4 The Tether Physics and Survivability Spacecraft (TiPS)

TiPS Program Overview

The Tether Physics and Survivability (TiPS) Experiment was conceived as a quick response, simple experiment to study the long term dynamics and survivability of tethered space systems. The knowledge gained from this experiment will help DOD and the nation gain experience with tethered systems for eventual use in operational spacecraft. The Naval Research Laboratory's, the Naval Center for Space Technology (NCST) designed, built and now operates the experiment for the National Reconnaissance Office (NRO). The experiment is a free flying satellite consisting of two end bodies connected by a 4 kilometer non-conducting tether. In this respect it is different from other tether experiments, like those flown on the Shuttle, where one endmass was connected to a massive host vehicle.

Figure 1.19 Artist Rendition of TiPS in Orbit. Ralph is on the bottom. TiPS has been in this orientation since deployment.

TiPS was jettisoned from a host spacecraft on June 20, 1996, with the deployment of the tether occurring shortly after jettison. The TiPS tether is intact through this writing (12/9/96), while no other space tether has lasted longer than five days. TiPS is the sixth known orbital tethered system flown to date.

To meet an early launch opportunity, TiPS had to be designed and built in approximately one year. Due to a very tight budget, the experiment objectives were limited to only those that had the highest payoff, these were : 1) Long term orbit and attitude dynamics and 2) tether survivability.

TiPS Hardware

The program constraints of limited time and money dictated that the experiment design be as simple as possible and consist largely of existing component hardware and/or designs.

The experiment goals only required a simple electrical power system that used a battery for all its electrical needs. The battery supported the initiation of deployment, recording of data on the deployment characteristics and transmitting this data to ground stations. TiPS consists of two end bodies, dubbed Ralph and Norton, connected by a 4 kilometer tether. Ralph contains a Small Expendable Deployer System (SEDS) tether deployer, a battery consisting of 10 Lithium Thionyl Chloride D cells, a timer to initiate deployment, a SEDS data acquisition electronics box and a transmitter and antennas to downlink the deployment data. The SEDS deployer, SEDS electronics and the transmitter were existing flight spare hardware from the SEDS 2 tether experiment that was successfully flown in space by NASA. NASA provided this hardware to NRL for the TiPS program. This was in keeping with the approach of using off the shelf components so that tight budgets and short schedules could be met. Norton is an inert body containing the ten spring cartridges used to rapidly push the two bodies away from each other, pulling the tether out of its deployer mounted on

Satellite Tracking and Dynamics

The motion of the end bodies is observed by a ground based Satellite Laser Ranging (SLR) network and by ground based visual observations. Fig. 1.21 is an image of TiPS taken as the two endmasses were separating. The tracking data consists largely of range data provided by timing the two way round trip delay for a laser bounced off the retroreflectors on the spacecraft. The data is transmitted across the Internet to NASA Goddard in Greenbelt, Maryland.

The tracking data is analyzed, at NRL, to determine the dynamic motion of the tethered system. Since, there is no object at the center of mass of the tethered system, both the attitude motion and the

Ralph. Ralph and Norton each have 18 optical retroreflectors mounted on them. Fig. 1.20 is a picture of the completed satellite without its thermal blankets. In this figure, several of the small round retroreflectors can be seen.

Fig 1.20 TiPS Satellite Without Thermal Blankets. The small round objects are the laser retroreflectors.

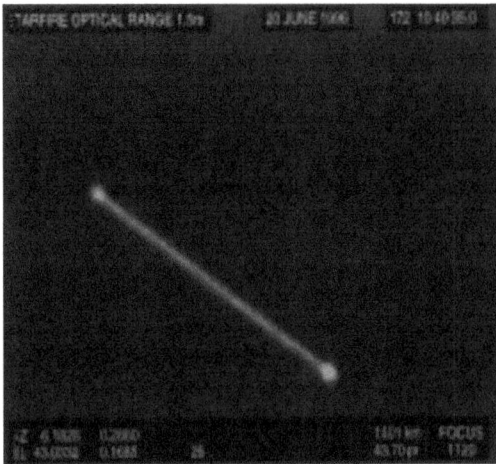

Figure 1.21 Telescope image of TiPS during deployment (Image taken by Starfire Optical Range at the Air Force's Phillips Laboratory).

26

orbital motion are inferred from observations of the endmasses only. This has proved to be a difficult task requiring frequent observational data and new estimation algorithms that were incorporated in the traditional orbit determination system used by NASA. The tracking data is also processed to provide updates to the state vector used to predict the motion of the endmasses. These predictions are used by the SLR sites for subsequent observations.

The requirement of determining the tether survivability will be met through ground based radar tracking of TiPS which will determine when or if the tether is cut.

TiPS Findings to Date

The findings of the TiPS program are best summarized by saying that they provide "confidence" that tether technology will be viable for future operational missions. There is still a lot of technology development required before operational systems would be ready to incorporate this technology, but TiPS has provided a significant step in that direction. The initial results show that tethers can be made to be survivable. With regard to the librational motion, our estimates now indicate that the tether is librating with a smaller amplitude than at deployment. Visual observations made shortly after the initial separation of the end-bodies suggested that the tether was librating with an amplitude of 47 degrees with respect to vertical alignment. Over the course of the next three months, we have determined with high confidence that the amplitude of that motion has decreased to approximately 12 degrees. At this lower amplitude, the tether behaves much more predictably. During the month of October, 1996, we were able to validate our ability to predict tether motion 6 to 12 hours into the future. While this was only possible during a period when an abundance of data is available, this provides a great deal of confidence in our ability to model tether dynamics.

NRL has set up a Web site where information and data can be obtained. The URL is http://hyperspace.nrl.navy.mil/tips.

Contacts for the Tips Project:

- Shannon L. Coffey, William E. Purdy, NRL

1.5 The OEDIPUS Tethered Sounding Rocket Missions

OEDIPUS A

OEDIPUS stands for **O**bservations of **E**lectric-field **D**istribution in the **I**onospheric **P**lasma - a **U**nique **S**trategy. Canadian activities in space tethers began with OEDIPUS A which was designed as a large double probe for sensitive measurements of weak electric fields in the plasma of the aurora. It was launched using a Black Brant X, 3-stage sounding rocket. The OEDIPUS program was a joint program between National Research Council of Canada and NASA with participation of the Communication Research Center in Ottawa, Canada (Principal Investigator), various Canadian universities, and the US Air Force Phillips Laboratory, the payload prime contractor was Bristol Aerospace Ltd. The major objectives of the OEDIPUS-A mission were:

- to make passive observations of auroral ionosphere, in particular, the natural magnetic-field-aligned dc electric field $E_\parallel$, utilizing a large double probe;
- to measure response of the large probe in the inospheric plasma;
- to seek new insights into plane- and sheath-wave rf propagation in plasma.

The rocket payload OEDIPUS A was flown on January 30, 1989 from Andøya in Norway. The tethered payload consisted of two spinning subpayloads with a mass of 84 and 131 kg, with their own experiment complement and telemetry systems, that were connected by a thin 0.85 mm diameter conductive and also spinning tether. The mission achieved its scientific objectives to detect the natural magnetic-field-aligned dc electric field $E_\parallel$ utilizing a large double probe, and to carry out novel bistatic propagation experiments. The flight established a record for the length of an electrodynamic tether in space at that time: 958 m. Although the mission was successful, flight data indicated that the aft subpayload experienced a rapid increase in its coning angle to nearly 35 degrees (half angle). A post-flight investigations concluded that the dynamic behavior was caused by interaction of the tether with the subpayloads. This observation was unexpected due the fact that the tether mass was negligible relative to masses of both subpayloads, and the tether dynamic interaction was expected to be negligible in the relatively short time (11 minutes) of a suborbital flight.

The OEDIPUS-A payload configuration is shown in Fig. *1.22*. The two subpayloads were initially connected and ejected from a Black Brant X with a spin rate about the longitudinal axis. The radial booms on the forward and aft payloads were used as dipoles for science experiments. The ACS module, located at the aft end of the aft subpayload, was used to align the spin axis to within 1 degree to the Earth's magnetic field. The tether was a teflon coated stranded tin-copper wire and it was deployed from a spool-type reel located on the forward subpayload. To separate the subpaylaods and deploy the tether, a spring ejection system was provided and followed by the cold gas thruster system in the forward subpayload. A magnetic hysteresis brake was provided to control the tether spool by applying a small constant torque, to smoothly decelerate the relative motion of subpayloads.

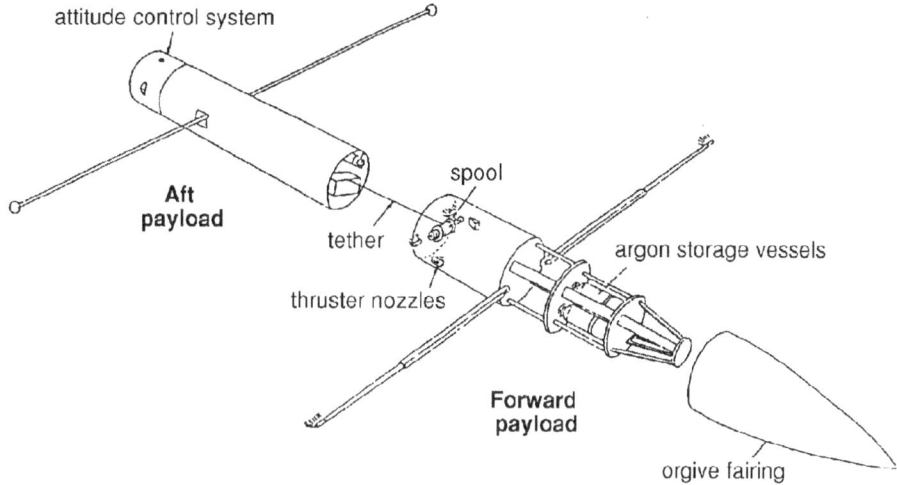

Figure 1.22 Some subsystems in the OEDIPUS-A payload configuration

Shortly after motor burn-out, the fairing was jettisoned along with a number of experiment doors, and the two sets of radial booms were deployed. At T+121 seconds, the ACS maneuver was initiated which aligned the payload within 1 degree of the local geomagnetic field line. At this point, the separation using cold gas system was initiated. At T+448 s, apogee was reached (about 512 km) and the payload separation was completed with tether length of 958 m. This configuration was maintained for the remainder of the flight. Due to gravity-gradient torque on the two-body system, the entire configuration experienced a slight rotation through-out the flight. At approximately T+800 s the payload re-entered the atmosphere and was not recovered.

Flight dynamics data are presented in Fig. *1.23*. These are processed magnetometer data used to compute the angular deviation of each payload's spin axis from the magnetic field vector. Data is shown for both forward and aft subpayloads. The forward subpayload experienced a tip-off at separation that caused a coning angle of approximately 7 degrees which varied only slightly during the flight. The aft subpayload, which also experienced a small tip-off, had an increase in the coning angle and it approached almost 35 degrees at the end of the flight. The post-flight investigations concluded that the increase coning was the interaction between tether and the aft subpayload.

OEDIPUS C

The second flight of OEDIPUS configuration, namely, OEDIPUS C took place on November 6, 1995 from the Poker Flat Research Range, located near Fairbanks, Alaska. The scientific objectives of the mission were similar to the previous one but there were

29

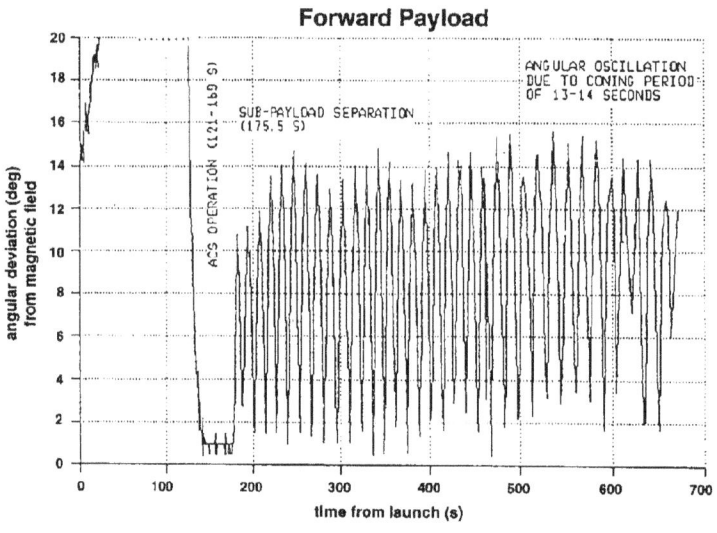

Forward Payload

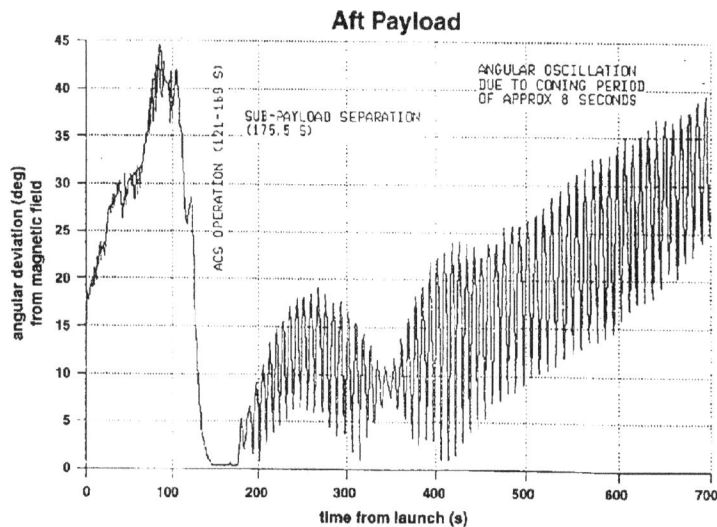

Aft Payload

Figure *1.23* OEDIPUS-A flight dynamics data showing the time history of the angle between spin axis and the direction of the earth's magnetic field, for both subpayloads

important differences and extensions. The OEDIPUS-C payload was launched, using the Black Brant XII Sounding Rocket, to a higher trajectory with apogee of 843 km and the length of deployed tether was 1174 m. Thus, the trajectory had a greater range in plasma density than OEDIPUS A and provided an extended perspective on plane and sheath waves and their interaction with space plasma. To understand the importance of the electrically conducting tether for the propagation of rf waves between the subpayloads, the tether was cut from both ends on the downleg part of the flight. The experiments were also designed to help understand how charged particles associated with the aurora affect satellite transmissions. There were 13 experiments (three instruments from the Canadian Space Agency, seven from the National Research Council of Canada, and three others from the University of Saskatchewan and the US Air Phillips Laboratory in the USA). The OEDIPUS-C payload was sponsored by the Canadian Space Agency and the payload contractor was Bristol Aerospace

30

Ltd. One of the main investigations on OEDIPUS C was the project funded by the CSA's Space Science Program which involved controlled radio-wave experiments. The radio instruments (HEX and REX) were built by CAL Corporation and Routes Inc., both from Ottawa, Ontario, and the Principal Investigator was from the Communications Research Center (CRC) in Ottawa, Canada.

The OEDIPUS-C configuration is presented on Fig. *1.24*. The tethered payload consisted of two spinning subpayloads with a mass of 115 and 93 kg, respectively. They were connected by the same type of the tether as used in the OEDIPUS-A mission (0.85 mm diameter). The subpayloads each had four long booms (Be-Cu BI-STEM elements), forming a V-dipole antenna (13 m, tip-to-tip, on the aft subpayload; and 19 m on the forward subpayload). The tether was a 24 gauge wire per MIL-22759/32 which has a 19 strand, tin-coated copper conductor with white teflon insulation (radiation cross-linked, modified ETFE) rated at 600 V. Both subpayloads had video cameras for the determining the and the attitude solution and the relative position between the forward and aft subpayloads. The payload performance was captured in space by an aft payload video camera. The subpayloads telemetered data to a ground station for about 15 minutes before they landed in the Arctic Ocean (non-retrievable).

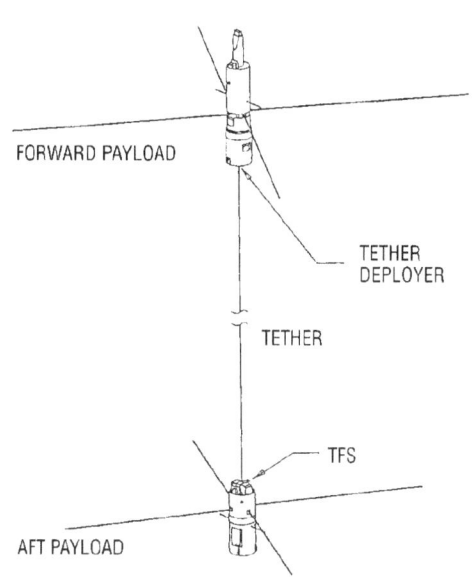

Figure 1.24 OEDIPUS-C configuration with location of the TFS

A unique Tether Dynamics Experiment (TDE) was one of the experiments flown during that mission. It was sponsored by the Space Technology Branch of the CSA in collaboration with Bristol Aerospace, University of Manitoba, University of British Columbia, McGill University, Carleton University and NASA Langley Research Center. A description of the TDE is presented in the following section.

OEDIPUS-C TETHER DYNAMICS EXPERIMENT (TDE)

The planning for this technological experiment was initiated in 1992, and culminated with the sub-orbital flight on November 6, 1995. The main objectives were as follows:

- derive theory and develop simulation and animation software for analyses of multi-body dynamics and control of the spinning tethered two-body configuration;

- provide dynamics and control expertise, for the suborbital tethered vehicle and for the science investigations, develop an attitude stabilization scheme for the payloads and support OEDIPUS C payload development;

- acquire dynamics data during flight, and compare with pre-flight simulations to demonstrate that the design technology is valid.

Figure 1.25 TEther LABoratory Demonstration System - TE-LAB, at DFL, CSA

Figure 1.26 Tether Force Sensor (TFS) flexure

presented in Fig. *1.27*.

The TDE advanced space tether technology significantly. The following are noteworthy.

- Several types of mathematical model were investigated, including both linear and non-linear approaches.
- A laboratory 'hanging spin test' facility was established at the University of British Columbia, which was able to demonstrate the essential dynamic stability characteristics of spinning tethered systems.
- A TEther LABoratory Demonstration System (TE-LAB), developed in conjunction with graduate engineering program of Carleton University, supported precise ground simulation of the OEDIPUS dynamics. The TE-LAB facility stimulated advances in gimbals suspension and in non-contact attitude measurement techniques, to meet stringent requirements of zero-g simulation in the one-g earth environment (Fig. *1.25*).
- A unique precision 3-axis tether Force Sensor (TFS) was designed by Bristol Aerospace Ltd. in conjunction with NASA Langley Research Center. The design derived from the NASA's experience with multicomponent wind tunnel balances for the aerospace industry. The TFS had two sets of strain gauges: foil gauges and piezo-resistive gauges. The TFS was manufactured by Bristol Aerospace Ltd., and Modern Machine & Tool Co., Newport News, Virginia, and was calibrated by NASA, CSA and Modern Machine & Tool Co. (Fig. *1.26*).

During the flight the subpayloads and all on-board instruments met and exceeded expectations. The deployment of the booms and tether, including severance of the tether from the payloads, was captured in space by the aft payload camera, and provided an overall confirmation of stability of the spinning subpayloads and tether dynamics. An example of the processed flight dynamics data - nutation angles of both subpayloads are

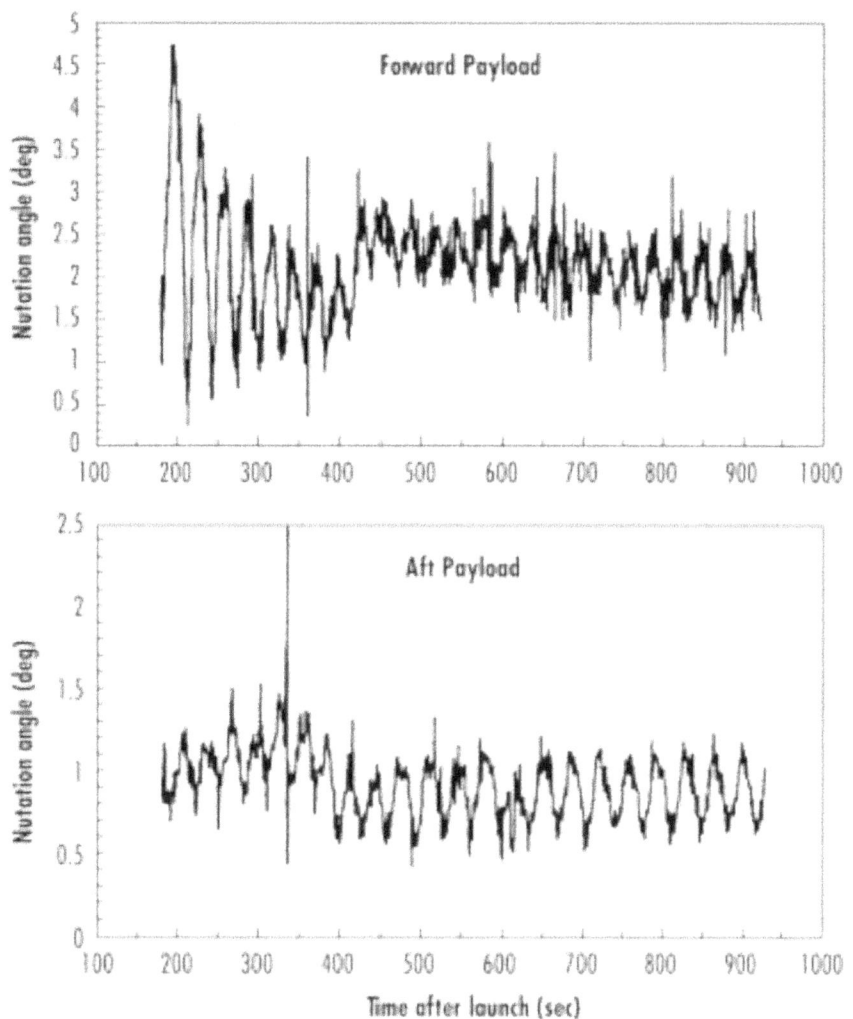

Figure 1.27 Nutation (coning) angles as function of time for OEDIPUS-C payload

The time history of the total tether force calculated based on the foil gauge outputs is presented in Fig. *1.28*. The deployment profile based on the flight data is showed in Fig. *1.29*. The major achievement was the implementation and demonstration of the major axis spinner stabilization for the tethered OEDIPUS-C subpayloads. The ground tests also served very well to understand the complicated dynamics of the spinning tethered two-body configuration and the interaction between the rigid and flexible body modes. The analysis of the damped gyroscopic modes of spinning tethered space vehicles with flexible booms turned to be a very effective tool to understand the dynamics of the *system.*

33

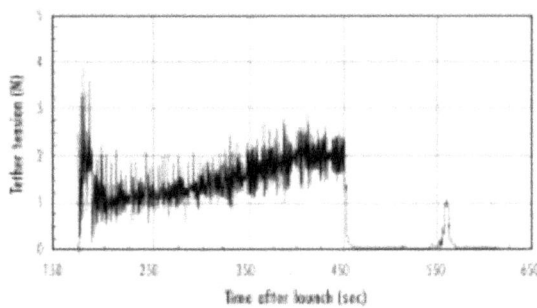

Figure 1.28 Tether tension vs. time measured by TFS during OEDIPUS C flight

The actual spin rate during flight was 0.084 Hz well within the stable range. Full 3-D computer animation of the tethered system's dynamic behaviour and of the damped gyroscopic modes also served very well in understanding the dynamics of this configuration. The OEDIPUS-C tether deployment system is presented in Figure 1.30. It was located in the aft end of the forward subpaylaod and it is comprised of a rotating spool, supporting structure, a magnetic hysteresis brake to control tether tension, a slip ring, high and low resolution shaft encoders, a wire guard/snare retainer, and forward tether cutter assembly.

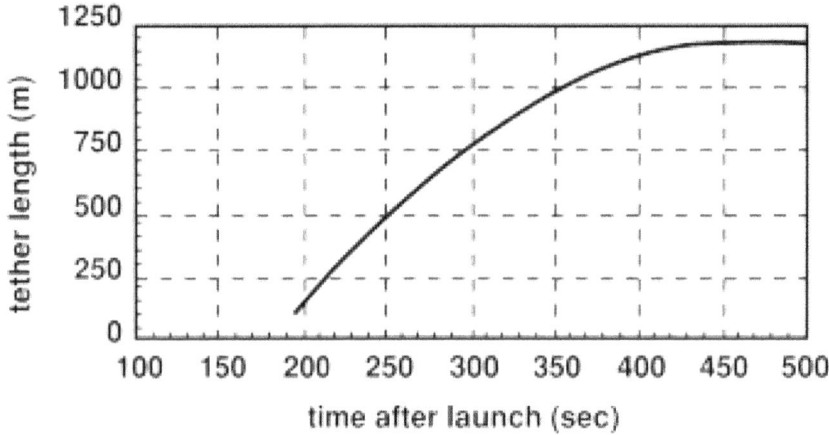

Figure 1.29 OEDIPUS-C tether deployment profile from the spool encoder data

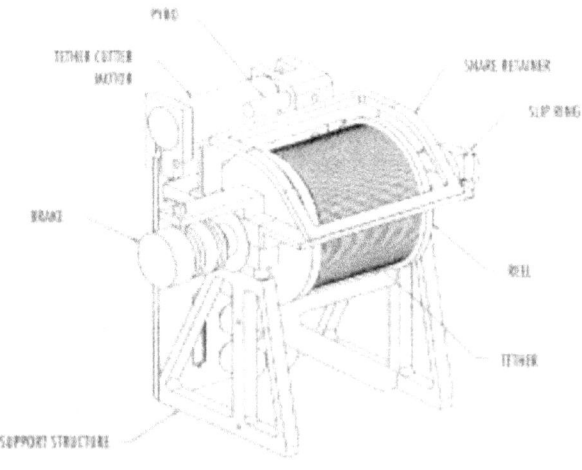

Figure 1.30
The Oedipus-C Tether Deployer

34

Contacts for the OEDIPUS Project:

- H. Gordon James - *Communications Research Centre*
- Alexander Jablonski - Canadian Space Agency
- George Tyc - Bristol Aerospace
- Frank Vigneron - Consultant, Canadian Space Agency

Contacts for the OEDIPUS-C Tether Dynamics Experiment (TDE):

- Alex Jablonski, Frank Vigneron - Canadian Space Agency
  Tether Dynamics Experiment, Payload Stabilizazion and TE-LAB
- George Tyc - Bristol Aerospace
  Tether Dynamics Experiment, Payload Stabilizazion
- Arun K. Misra - McGill University
  Tether Dynamics
- Vinod J. Modi - University of British Columbia
  Tether Dynamics
- Douglas A. Staley - Carleton University
  TE-LAB
- Ray Rhew - NASA LaRC
  TFS

SECTION 2.0 PROPOSED TETHER FLIGHTS

2.1 Electrodynamic Tethers For Reboost of the International Space Station

Propellantless Reboost for the ISS: An Electrodynamic Tether Thruster

The need for an alternative to chemical thruster reboost of the ISS has become increasingly apparent as the station nears completion. A new type of electrodynamic tether attached to the Station (Figure 1) could be developed to generate an average thrust of 0.5-0.8 Newtons for 5-10 kW of electrical power. By comparison, aerodynamic drag on ISS is expected to average from 0.3 to 1.1N (depending upon the year).

The proposed system uses a tether with a kilometers-long uninsulated (bare) segment capable of collecting currents greater than 10 A from the ionosphere. The new design exhibits a remarkable insensitivity to electron density variations, allowing it to operate efficiently even at night. A relatively short and light tether (10 km or less, 200 kg) is required, thus minimizing the impact on the ISS (center of mass shift less than 5 m).

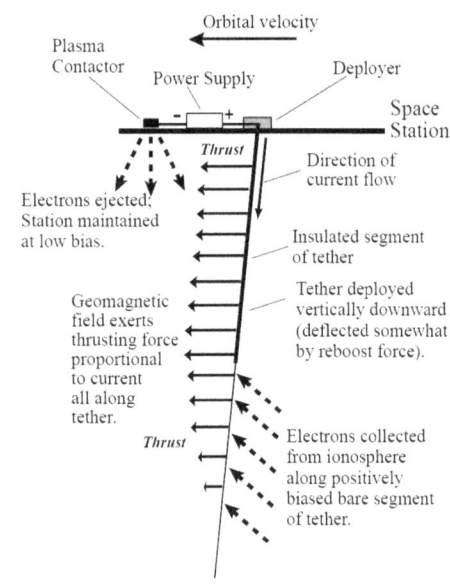

Figure 1. An electrodynamic tether reboost system for the International Space Station.

High Tether Currents for ISS Reboost

ISS reboost (thrust forces of order 1 N) with a tether no longer than 10 km requires tether currents of order 10 A. The critical issue is how to draw ionospheric electrons at that rate. The standard tether carries insulation along its entire length, exchanging current with the ionosphere only at the ends: TSS-1R carried a passive metallic sphere as anode; PMG carried an active (plasma-ejecting) contactor.

Current collected to a passive, biased sphere in a magnetized plasma calculated by the standard Parker-Murphy (PM) model (taking into account magnetic effects, which are dominant) grows as the square-root of the bias voltage, an important fact for fixed-area collectors.

A preliminary analysis of the measured TSS-1R currents indicates that they were typically greater than the PM model predictions (using values of the electron density and temperature estimated from ionospheric models and a satellite voltage calculated with some uncertainty). The TSS-1R data do not, however, appear to point to a dependence of current on voltage greatly different from that of PM for higher voltages. Even though, for example, a TSS-1R current of 0.5A at 350 V bias may surpass PM model estimates, it could still imply a voltage of roughly 35 kV to reach 5 A for the same plasma parameters (which would require over 175 kW for a thrust of 0.7 N with a 10-km-long tether!).

Active anodes (plasma contactors) have been developed in an attempt to solve both space-charge shielding and magnetic guiding effects by creating a self-regulating plasma cloud to provide quasineutrality and by emitting ions to counterstream attracted electrons and produce fluctuations that scatter those electrons off magnetic field lines. The only tether experiment to use an active anode so far was the PMG, which reached 0.3A in flight under a 130 V bias and the best ionospheric conditions. Unfortunately, there is no way to scale the results to high currents. The discouraging fact was that collected current decreased sharply with the ambient electron density at night.

Fortunately, there is another tether design option— the bare tether - as proposed by Sanmartín .

<u>The Bare-Tether Breakthrough</u>. The bare-tether design represents a breakthrough that makes short-tether electrodynamic reboost with moderate power requirements for the ISS a possibility. To work on the ISS, a reboost system must not only be capable of delivering adequate thrust (preferably night and day); it must do so with small impact on the ISS environment while requiring minimal accommodation by the baseline ISS systems. It should also be simple to operate and maintain, and it must be competitive in terms of its use of resources for the benefits it provides.

Our proposed design uses the tether itself, left uninsulated over the lower portion, to function as its own very efficient anode. The tether is biased positively with respect to the plasma along some or all of its length. The positively biased, uninsulated part of the tether then collects electrons from the plasma.

The following features argue in favor of the bare-tether concept.

1. The small cross-sectional dimension of the tether makes it a much more effective collector of electrons (per unit area) from the space plasma than is a large sphere (such as the TSS-1R satellite) at equal bias. This is because the small cross dimension of the tether allows its current collection to take place in the orbital-motion-limited regime, which gives the highest possible current density.

2. The large current-collection area is distributed along the tether itself, eliminating the need for a large, massive and/or high-drag sphere or a resource-using plasma contactor at the upper end of the tether. This substantially reduces the center of gravity shift in both cases and reduces the cost and complexity in the case of the active contactor.

3. The system is self-adjusting to changes in electron density. This is accomplished by a natural expansion of the portion of the tether that is biased positively relative to the ionosphere whenever the density drops (Figure 2).

Features (1) and (2) combine to provide an ability to collect large currents with modest input power levels. We present below a candidate system that can produce average thrusts of 0.5-0.8 N, for input power of 5-10 kW.

<u>Developing an ISS Reboost System</u>. Our preliminary design for an electrodynamic tether thruster capable of delivering 0.5-0.8 N of thrust to the ISS at a cost of 5-10 kW of electrical power consists of an 10-km-long aluminum tether in the form of a thick ribbon (0.6 mm by 10 mm). Despite its length, the tether would weigh only around 200 kg. Since the bare portion of the tether is to act as our electron collector, a downward deployment of the tether is dictated by the physics of the eastward-moving platform.

The upper part of the tether will be insulated. There are two reasons for this. First, there is the necessity for preventing electrical contact from developing across the plasma between the upper portion of the tether and the Space Station, which (when the system is operating) are separated by an electrical potential difference of around a kilovolt. Beyond that, the insulation provides for greater thrust at a given input power. This comes from the fact that the largest tether-to-plasma bias occurs at the upper end, and decreases down the tether. A completely bare tether would draw the maximum

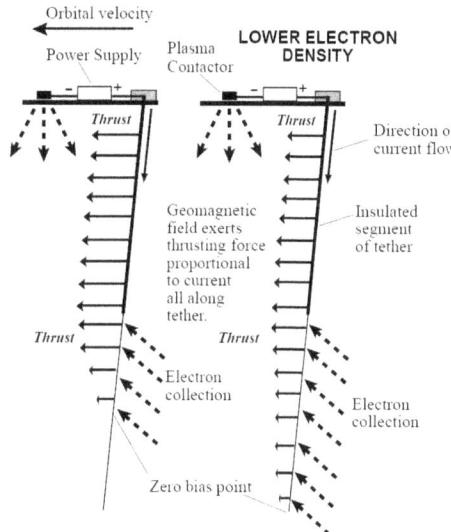

Figure 2. A bare-tether thruster designed to adjust to lower electron density (as at night). A shift in the zero point of bias further down the tether increases the collecting surface and maintains a nearly steady thrust for constant input power and induced e.m.f.

current through the power supply, but the current would be strongly peaked at the upper end of the tether. Keeping the input power constant, we can substantially increase the average current in the tether, and hence the thrust, by insulating the tether over much of its upper portion, collecting current with the lower portion, and having a constant current in the upper part.

Determining the optimal fraction to insulate is part of the design effort for a "bare" tether reboost system. Our preliminary design has the upper 50% of the tether insulated. Even greater thrust during daytime operation could be obtained with a higher fraction, but the night–time adjustability would suffer.

The system provides flexibility, in the sense that the thrust obtained depends almost linearly on the input power, as seen in Figure 4.

The bare-tether design has essentially solved the problem of day/night thrust fluctuations. But fluctuations in thrust due to fluctuations in the induced e.m.f. as the system encounters a varying geomagnetic field around the orbit are a fact of life for any tether-based system. Figure 5 show the thrust variations around the ISS orbit with different input power levels.

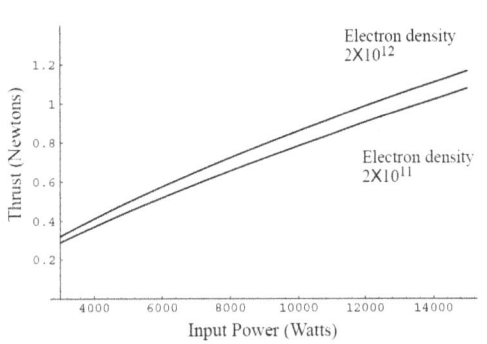

Figure 4. Variation of thrust with input power for nominal 10-km system. Motional e.m.f.: 1.2 kV.

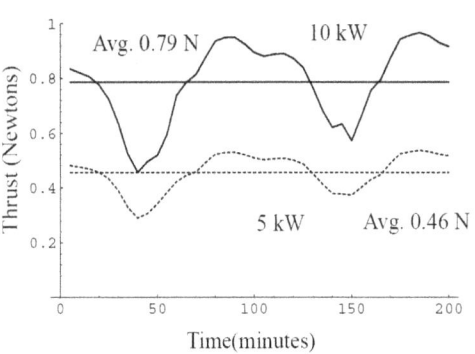

Figure 5. Comparison of thrust generated for input powers of 5 kW and 10 kW

Given the level of the current the system may draw, the system will almost certainly require its own cathodic plasma contactor at the Station end. The contactors currently under development at NASA Lewis Research Center should be well suited for this function. If thrusts over 0.5 N are desired, it is likely that the system will also have to rely on the ISS's plasma contactor as well, or on a second dedicated contactor, since currents over the 10 A rating of the contactors could be required.

Before an operational electrodynamic tether reboost system for the ISS can be designed, a series of ground and space-borne experiments and computer simulations must be performed. In addition, thorough systems analyses must be performed to determine the physical integration and operational issues associated with its implementation on the ISS.

Among the issues to be addressed in the analyses of the reboost system are the attachment location for the tether, need for retrieval capability, microgravity impact, power interfacing, and safety. These are in addition to design issues specific to the tether itself, such as tether material, length, and geometry.

Assessment of Space Application and Benefits to the ISS

1) Mission Benefit. The value in an electrodynamic tether reboost system lies in its ability to couple power generation with thrust. Heretofore the electrical and propulsion systems have been effectively totally separate entities. Outfitting ISS with an electrodynamic reboost tether severs the most critical and constraining dependency on Earth - propellant resupply. The Station can supply its own power but not its own propellant. Without an electrodynamic tether, the specter of SkyLab and the

words "reentry" and "atmospheric burnup" will forever haunt the minds of anyone who has an interest in the program. Add a tether and some additional storage capacity for supplies, and suddenly a one year interval between visits to the Station becomes conceivable.

Even if the current frequency of resupply flights to the Station is maintained, with an electrodynamic tether the Station Program has the option to trade kilowatts for increased payload capacity. Resupply vehicles can deliver useful cargo like payloads, replacement parts, and crew supplies rather than propellant. Within the range of 5 to 10 kW, a crude approximation of 1,000 kg of user payload gained per kW expended per year appears reasonable; further analysis will refine this estimate.

As a bonus, propellantless reboost is exhaustless reboost: external contamination around the Station is considerably reduced. The Station reboost propellant is hydrazine. Any consumption of propellant may result in residual chemical deposits and contamination on the Station's exterior surface. An electrodynamic tether provides a means to reboost the Station without the complications of chemical combustion. The purity of the external environment for science payloads is enhanced, and beneficial operational impacts of reduced propellant exhaust on external systems and optics will be realized. Electrodynamic thrust truly represents solar power at its finest.

Yet another dimension to propellantless reboost must be considered. Station users have been allocated a minimum of 180 days of microgravity per year. Current planning essentially halts science activity during reboost maneuvers. Low thrust electrodynamic tether reboost could be performed over long duration, as opposed to short duration, high thrust propulsive maneuvers. The 0.5 to 0.8 N thrust provided by a 10 km tether more than counteracts the Station's atmospheric drag on a daily basis. Thus the question arises, can an electrodynamic tether compensate for the drag while it is occurring, without disrupting the microgravity environment? Fluctuations in the induced voltages from the Earth's magnetic field and in electron densities will create "turbulence" through which the electrodynamic tether driven Station must fly; can load-leveling control systems compensate for these pockets and maintain microgravity levels? In this case a new realm of possibilities opens up for long-duration microgravity experiments. The allure of this self-propelled space facility is certainly remarkable, and offers potential advantages.

2) Risk Reduction. Aside from replacement of failed components, an electrodynamic reboost tether on the Station makes the vehicle itself essentially independent of propellant resupply from Earth. The primary resupply consideration becomes the inhabitants of the Station and not the Station itself. This is a new view for development of space operations. There ceases to be concern over the "180-day countdown to reentry at 150 nautical miles" which currently permeates every aspect of Station mission planning. With the multi-billion dollar investment in the vehicle virtually secured and free from concern over long resupply vehicle launch delays, particularly Russian Progress or FGB tanker delays, the Program will be able to focus much more strongly on the ISS mission rather than on ISS itself.

3) Cost Pay Back. The cost of the proposed system comes in the form of the development, launch, and installation of an operational tether reboost system on the Station. The payback comes in the form of reduced propellant upmass requirement. For 2003 to 2012, nearly 90,000 kg of propellant must be launched. Using a figure of $20,000 per kg, this represents a sum of $1.8 billion. An electrodynamic tether supplying 90 percent of this requirement would reduce the operational cost by $1.6 billion, paying for itself many times over. More modest estimates still result in a return on investment tens of times the cost of development and operation of an electrodynamic reboost tether.

Contacts:
Les Johnson, NASA/MSFC
Joe Carroll, Tether Applications Company
Juan Sanmartin, Polytechnic University of Madrid
Robert D. Estes & EnricoLorenzini, Smithsonian Astrophysical Observatory
Brian Gilchrist, The University of Michigan Ann Arbor
Manuel Martinez-Sanchez, Massachusetts Institute of Technology

2.2 An Upper Atmospheric Tether Mission (ATM)

Introduction

The Atmospheric Tether Mission (ATM) is a Shuttle based scientific experiment that will deploy a set of eleven instruments to collect valuable atmospheric data never before obtained. This set of instruments will be housed in an endmass/spacecraft that is deployed downward from the Shuttle by a 90 km tether. The instrument package will cut through the atmosphere, collecting data, at three different altitudes over a six day mission. A team was formed at the Marshall Space Flight Center (MSFC) to conduct a preliminary concept study defining a system that would accomplish the objectives of the ATM. A detailed report will be published by the team at the conclusion of the study.

Science Instrument Requirements

A Science Definition Team (SDT) was formed by NASA Headquarters to define the scientific objectives of the Atmospheric Tether Mission (ATM). The SDT proposed a set of eleven science instruments that together would meet all of the ATM mission objectives. The instruments, their requirements and locations are shown in Table 1 and in Fig. 6, respectively.

Instrument Description	Sensor Dimensions (cm)	Electronics Dimensions (cm)	Sensor Mass (Kg)	E-Box Mass (Kg)	Instrument Power (W)	Telemetry Rate (bps)
Ion Drift Meter	12 dia 7 deep	21x12x16	0.9	2.3	3	2000
Retarding Potential Analyzer	12 dia 7 deep	21x12x16	0.9	2.3	4	1000
Ion Mass Spectrometer	18x12x11	18x12x16	1.8	2.0	6	500
Langmuir Probe	1 dia 15 long boom mount	15x15x10	0.35	3.0	4	5600
Neutral Wind Meter	16 dia 19 deep	18x12x16	2.1	2.2	8	1000
Neutral Mass Spectrometer	18x12x11	18x12x16	2.0	2.5	10	1000
Energetic Particle Spectrometer	19x15x18	Included in Sensor	2.2	N/A	2	8000
E-Field Double Probes	20 cm dia 6 deep	12x12x8	18.0 (3x6)	3.0	10	50K
IR Spectrometer	10x10x21	18x18x13	7.0	2.0	13	128K
UV Photometer	10x10x25	inc. in sensor	2.8	inc. sens	5	320
3-Axis Magnetometer	8x8x21	18x18x13	1.0	2.5	2	1600
Total Payload			39.1	21.8	67	199K

Table 1. Science Instrument Requirements

Figure 6. Preliminary drawing of the ATM endmass

Mission Scenario

The baseline mission scenario is that the Orbiter will enter a 220 km circular orbit at a 57 degree inclination. The tether length for this scenario is 90 km and will operate in a deploy only mode. On the first day the tethered endmass will be deployed downward 50 km to 170 km altitude and remain there for two days. On day three, an additional 20 km will be deployed, lowering the endmass to an altitude of 150 km for two days. On day five, the final 20 km of tether will be deployed, lowering the endmass to its final 130 km altitude and will remain at this altitude for two days. The Orbiter altitude will be maintained by use of the Primary Reaction Control System (PRCS) thrusters on the Orbiter. On day seven, the tether is cut and the endmass begins a reentry course. The current estimate of fuel required for this scenario is 1996 kg (4400 lb.).

Five of the science instruments are required to face the RAM direction with two in the wake. A series of E-field double probes and Langmuir probes are placed at specific locations around the 1.6 m diameter satellite shell. This concept shows an aerodynamic tail used to increase yaw stability.

Aerodynamic Analysis of Endmass

The drag for a spherical shaped endmass of 1.6 m in diameter ranges from 0.92N at 130 km altitude to 0.11N at 170 km altitude. A bullet shaped endmass was considered to ease packaging constraints of the endmass subsystems. The drag analysis showed that the drag for a sphere is 20 percent lower than the equivalent bullet shape endmass. The diameter of the spherical endmass was increased from 1 m to 1.6 m in diameter to alleviate packaging constraints.

Endmass Attitude Control System

There are several constraints impacting the endmass attitude control system design. Two major constraints on the system are; avoidance of large torques that will disturb the endmass force and acceleration measurements, and the inability of using magnetic torquers because they cause disturbances in the magnetic field flux measurements. The science instrument requirements state that the endmass should be pointed within plus or minus 3 degrees of RAM with a plus or minus 0.1 degree post-flight knowledge requirement. An attitude control system combining the use of reaction wheels and strategically placed cold gas thrusters is the current proposed baseline. The location of thrusters will be determined using Direct Simulation Monte Carlo (DSMC) analysis to avoid instrument and endmass contamination. The control system is estimated to weigh 15 kg.

Electrical Power System

The mission lifetime of six days requires seven Li/SOCL2 type batteries weighing 105 kg. The additional cables, harnesses and distribution weights bring the electrical power system to an estimated 155 kg. The total desired power loads are estimated at 176.6 watts including a 25 percent contingency. This total includes the science instruments and electronics, and the endmass major subsystem equipment. A summary of the electrical power system mass versus mission durations is seen in Figure 7.

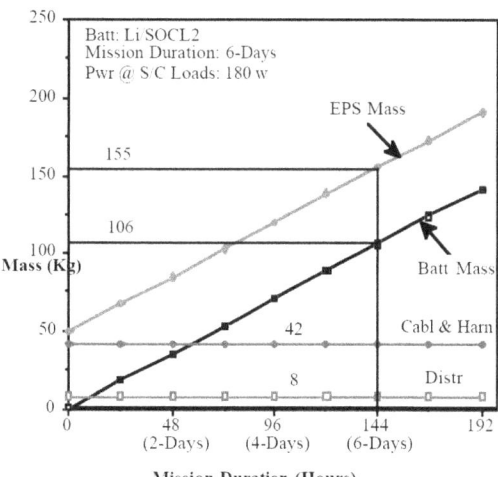

Fig. 7. ATM Electrical Power System Mass

43

Thermal Control System

A flowfield temperature analysis was performed at an altitude of 130 km. The temperature variations occur in shock-layers ranging from 800 K to 12000 K. The maximum aero-heating on the endmass surface is shown in Figure 8.

A combination of thermal blankets and heaters comprise the current endmass thermal control system. The estimated weight of the system is 7 kg requiring 4 W of power.

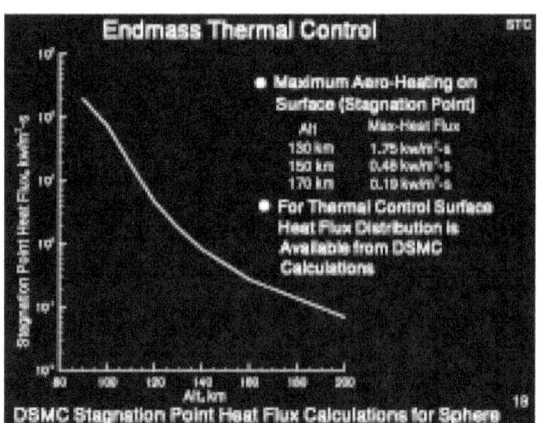

Fig. 8. ATM endmass thermal control

Endmass Structure

The recommended material for the endmass structure is Aluminum 2219. The endmass structure is composed of an equatorial ring with a mounting panel with two hemispheres of four flanged quadrants each. Local stiffening will be required for the mounting of deployables and some instruments, and attachment of the aerodynamic tail. A smooth surface is desired for aerodynamics requiring the use of closeouts. The estimated weight of the endmass structure is 81.9 kg.

Baseline Tether Concept

The current tether concept is a 1.65 mm diameter Kevlar strength member surrounded by a Nomex jacket with a total diameter of 2.16 mm. A magnification of the baseline tether is shown in Figure 9.

The tether has a break strength of 2892 N and weighs 4.03 kg per km. The tether is non-conducting and is currently 90 km in length. The probability of survival of the baseline tether over a six day mission, assuming a critical particle size of 0.3 of the tether diameter, is approximately 0.93. The probability of survival is highly sensitive to critical particle size. A graph showing a particle size of 0.2, 0.3, and 0.5 of the tether diameter is seen in Figure 10.

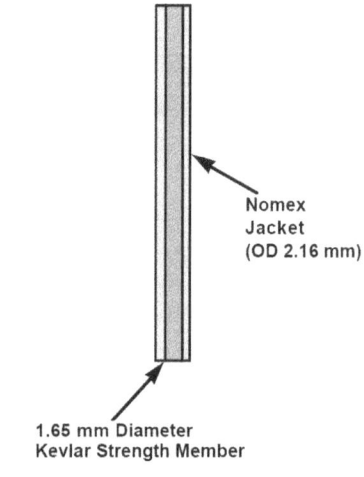

Fig. 9. A magnification of the baseline tether.

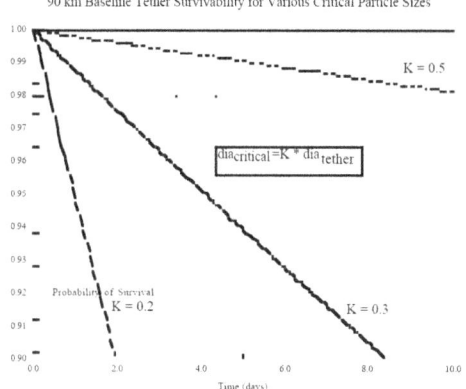

Fig. 10. Comparison of critical particle sizes and the probability of survival versus time.

Several alternate tether designs are being considered like the Hoytape (see *Failsafe Multiline Tethers for Long Tether Lifetimes* in the Application section). The survival probability using a particle size of 0.3 of the tether diameter jumps from 91 percent for a single line tether to 99.99 percent for the Hoytape. With a Hoytape type of tether, there is increased surface area increasing the overall drag on the tethered system. Other Hoytape designs using smaller diameter members will improve the drag concern while maintaining a near 100 percent survival.

Atmospheric Drag and Tether Dynamics

The atmospheric drag on the tether and endmass will induce libration oscillations of the tether. This is due to the fact that the atmospheric density is not constant thus affecting the in-plane libration of the tether.
Based on the current analysis, a libration and/or satellite pitch attitude control scenario may be required.

Deployment Dynamics

There are two types of deployers considered for the ATM mission. A modified Tethered Satellite System (TSS) deployer and a Small Expendable Deployer System (SEDS). The TSS deployer exists and has flown twice but must be modified for the ATM mission. The SEDS deployer is smaller but is not Orbiter qualified and would require extensive modification. The current baseline deployer of the ATM system is a modified TSS type deployer. Deployment dynamics are stable and have been demonstrated in earlier missions. The TSS deployment control strategy is proven and suitable for the expected endmass altitudes required in the ATM mission. The proposed ATM system will be mounted on a Spacelab Pallet in a designated location in the Orbiter payload bay.

Weight Statement

The total estimated weight (without contingency) of the endmass is 325.3 kg. The deployer reel, electronics, support structure and Spacelab pallet add an additional 2940 kg and the tether adds 500 kg. With a 30 percent contingency the total weight of the ATM system is 4895 kg. Table 2 details the ATM weight statement.

Table 2. ATM weight statement.

- Endmass
  - Science Instruments & Electronics Boxes — 60.9 kg
  - Structures — 81.9 kg
  - Electrical Power System — 155.0 kg
  - C&DH System — 5.5 kg
  - Thermal Control — 7.0 kg
  - Attitude Control System — 15.0 kg
- Deployer
  - Reel, Electronics, Support Structure, SLPallet — 2940.0 kg
  - Tether (120 km) — 500.0 kg
- Contingency (30%) — 1129.6 kg
- Total — 4894.9 kg

45

ATM Development Schedule

From Authority To Proceed (ATP), the development of the ATM is planned to take four years. A six month Phase A study for engineering design would begin immediately followed by a nine month Phase B definition. Parallel to the beginning of the Phase A, an Announcement of Opportunity (AO) would be released for the science instruments. The selection of the instruments would occur at the beginning of the Phase B and the science instrument design, development, fabrication and testing would begin. The development of the endmass and tether would begin parallel to the instrument development with the deployer development starting within the next quarter. All hardware would be delivered and integrated into the Orbiter in the beginning of the fourth year with a projected launch in the third quarter of the year.

Contacts:
- Les Johnson, NASA-MSFC
- B. Carovillano, T. Stuart, NASA-Headquarters
- R. Heelis, U. Texas

2.3 The Naval Research Laboratory's Advanced Tether Experiment

The Naval Research Laboratory (NRL) plans to fly its second tether experiment, called ATEx, in 1998. ATEx stands for Advanced Tether Experiment. The tether system is a simple gravity-gradient dynamics and survivability research experiment.

Major program objectives include adding to the tether community's understanding of deployment dynamics and control via a constant-speed motor, in- and out-of-orbit plane libration control via thrusters to excite and damp librations, and investigating the survivability of long-life tether materials.

Isometric Views of ATEx

Upper End-Body

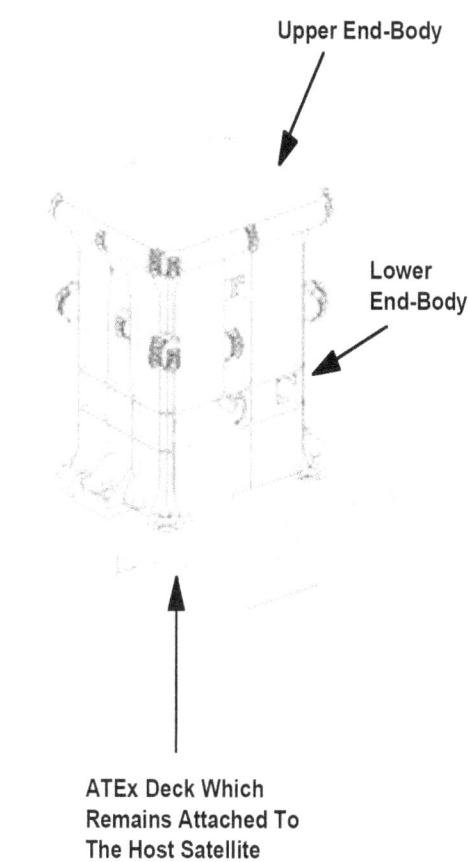

Lower End-Body

ATEx Deck Which Remains Attached To The Host Satellite

```
Upper End-Body:  7.6 x 62.2 x 52.0 cm
Lower End-Body: 60.9 x 48.2 x 38.1 cm
```

<u>Mechanical Overview</u>

The 83 kg tether system will fly as a payload on a host satellite in a circular altitude of 425 NM. A passive upper end-body's mass of about 12 kg has no instrumentation other than green-filtered retroreflectors. A 6 km (12 kg) tether is composed of 0.004 inch thick by 1 inch wide low density polyethylene with 3 single strands of 215 denier Spectra® 1000 uniformly spaced across the width. The lower end-body, of 30 kg mass, remains attached to a 29 kg electronics deck for the 90-day attached phase of the mission. At the end of the 90-day tether experiment, the lower end-body is separated from the electronics deck, which remains with the satellite. The lower end-body and portions of the satellite are covered with IR-filtered retroreflectors.

To accomplish some of the mission's science objectives, the lower end-body is instrumented with a 3-axis tensiometer at the tether attach point, a 3-axis accelerometer, a reel turn-counter, and a sensor to detect some discrete angles of tether departure with respect to the lower end-body.

Deployment Scenario

After achieving a near-circular orbit, the 3-axis stabilized momentum-bias satellite will orient with ATEx radially away from Earth. ATEx's upper end-body will separate away from the lower end-body via a constant speed motor at 2 cm/s. The stepper-motor will drive a pair of pinch rollers pulling the tether off a level-wound reel; but, the motor and reel cannot reverse direction. The entire deployment sequence has been specified and includes satellite pitch motions to maintain a tether departure angle nearly perpendicular to the lower end-body.

Analysis showed the in-plane system libration angle will initially be fifty degrees and throughout the deployment oscillate at significantly lower angles to result in a final libration angle near zero degrees.

Libration Control Demonstrations

For the remaining 87 days of post-deployment activities, tether dynamics will focus on exciting and damping in- and out-of-plane librations. The satellite has thrusters located on all four sides of the vehicle to force the satellite and lower end-body (now acting as one large end-body) forward-and-back in the orbit plane and left-and-right out the orbit plane. Details of these activities have not been defined; however, a thruster would be fired and observations made of tension, acceleration, satellite attitude perturbations, and end-body positions. The results would be interpreted in a quick-look scheme via the dynamics simulations.

Satellite Laser Ranging (SLR) Tracking

Each end-body has 43 retroreflector optics or "corner cubes". A retroreflector returns light back to the source independent of retroreflector orientation thus permitting the end-bodies to be observed by the global SLR network. The different coating on each end-body is sensitive to a different laser frequency to assist in identifying the end-body. Early in the mission, telescope observations will guide the laser beam to the end-body.

Later in the mission, perhaps the tether motions will repeat regularly and orbit determination will be straightforward such that a laser can target the end-bodies even in local daylight.

The SLR ground stations require pointing information given by inter-range vectors (IRVs). NRL will enhance the tether system's orbit determination from USSPACECOMMAND by including end-body motions. Initially, the tether dynamics models of the in-plane and out-of-plane librations will be used to augment the IRV.

Later, as SLR data becomes routinely available, estimates of the orbit and refined tether dynamics models from the SLR data should substantially improve end-body position and rate estimates. The IRV can be fit to the observed tether dynamics to enhance the acquisition and tracking, perhaps the SLR sites can acquire (in daytime) without telescope assist. This will increase around the globe viewing opportunities.

The Goddard Space Flight Center coordinates SLR observations within their international network and distributes the IRVs to each site. We expect to collect tether data for approximately one year. After that, we plan to occasionally request a series of SLR tracks to confirm long-term tether motion and that the tether is still intact.

Tether Survivability

After the 87 days of libration control research, the lower end-body is separated from the satellite. At this time, ATEx is completely unpowered, passive, and can only be observed by the ground methods: SLR, radar, optical telescops. At this point, the ATEx mission is similar to the TiPS project described in chapter 1 of this handbook. Analyses indicate that ATEx will reenter into the atmosphere in 3-4 years. The model included the atmospheric heating effects of the solar cycle.

Contacts:

- D. Spencer, M. F. Zedd - NRL

2.4 The AIRSEDS-S Mission

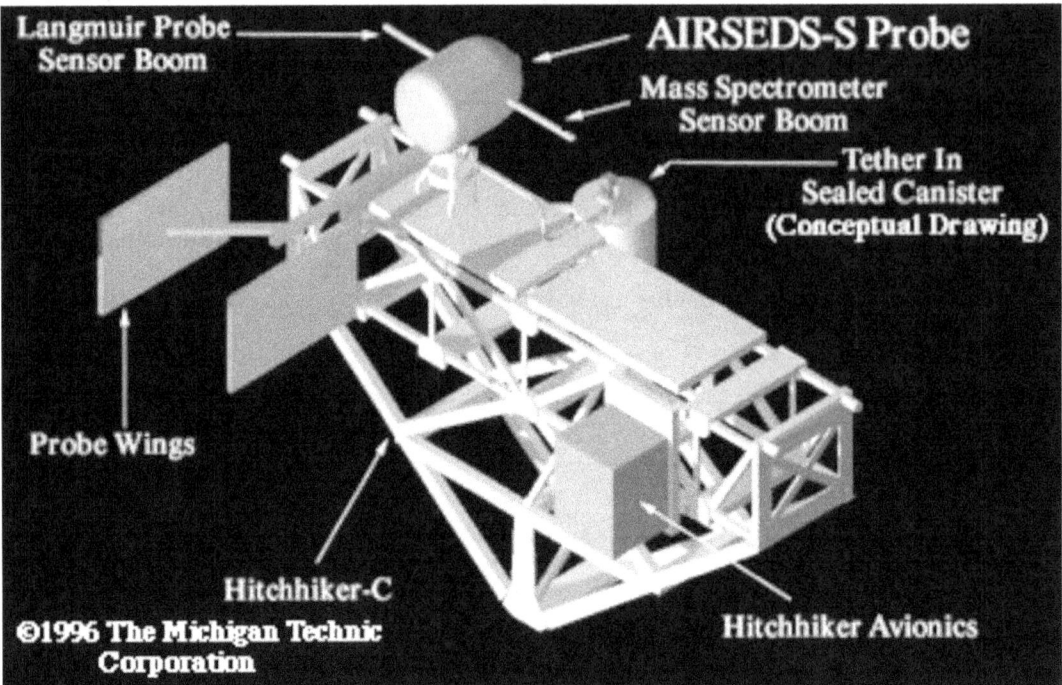

Several organizations have expressed a need for low cost tether solutions for the space shuttle, International Space Station and unmanned launch vehicles. Furthermore, NASA and ASI have expressed interest in flying a tethered satellite system in a downward deployed mission called TSS-2.[1] The scientific and engineering information to be gained from such a mission would allow advances in our understanding and modeling capabilities of atmospheric and ionospheric phenomena including satellite drag, the energy deposition from magnetospheric currents and particle precipitation, and the spatial and temporal gradients in ionospheric properties. Moreover, the next generation of tethered satellites and hypersonic vehicles are being planned to fly through this atmospheric region. Before undertaking a mission of the size and complexity of TSS-2 it may be prudent to explore the possibility that a less complex mission might be performed, which utilizes many present tether technologies and options for commercial sponsorship, to achieve a limited set of science and engineering goals. In the fall of 1994 The Michigan Technic Corporation (TMTC) was awarded by NASA Headquarters and Marshall Space Flight Center Phase A funding to conduct a preliminary design of the AIRSEDS-S probe and mission plan. AIRSEDS-S, Atmospheric/Ionospheric Research Small Expendable Deployed Satellite, will test and demonstrate tether system dynamical interactions, flight qualify deployer systems and reusable components for application to the Space Shuttle and the International Space Station (ISS), verify models of tether and satellite aerothermodynamic behavior, and determine lower thermosphere chemistry and composition. Figure 1 shows the system conceptually integrated with the Hitchhiker-C Crossbay Structure. AIRSEDS-S, based on NASA's successful and proven SEDS program, is a 90 km tether mission designed to collect atmospheric information in the altitude range of 230-130 km via a tethered satellite lowered from the Space Shuttle Orbiter to altitudes which cannot currently be explored using balloons or aircraft. The AIRSEDS-S mission will provide the first horizontal in-situ sampling at low altitudes in the Earth's upper atmosphere. In addition, the successful flight demonstration of the AIRSEDS-S probe and deployer system could result in the future development of a low

50

cost deployer to conduct further exploration of the Earth's upper atmosphere and ionosphere, and conduct payload return operations, ISS towing operations and microgravity experiments from the space shuttle and the International Space Station.

The specific objectives of the AIRSEDS-S Mission are to:

(a) Flight qualify tethered satellite hardware on the Space Shuttle Hitchhiker-C and for use, by inference, on the International Space Station.

(b) Conduct an investigation of the horizontal distribution of neutral atmosphere composition and dynamics in the lower thermosphere.

(c) Understand the local atmospheric environment of the tethered probe, and compare with current predictions.

(d) Test and demonstrate tether system dynamical interactions. This includes studying the behavior of a tethered satellite system and analyzing the flight characteristics of the probe in the Earth's upper atmosphere and comparing with current models.

(e) To provide educational opportunities to students in both pre-college and college level.

The long term goal of TMTC and the participants of the AIRSEDS-S mission including the University of Texas at Dallas, the University of Iowa, the University of New Hampshire, The AIRSEDS Institute, Tether Applications, Tethers Unlimited, The Smithsonian Astrophysical Observatory, and NASA Marshall and Goddard Space Flight Centers, is to provide a low-cost reusable modular tether facility for the International Space Station (ISS) and the Space Shuttle Hitchhiker-C programs. Such a facility may be further developed to support experiments conducting remote sensing, electrodynamic operations, microgravity studies and payload return. Most of the components for the AIRSEDS-S mission will have direct application on future ISS applications and shuttle based missions including the deployer system, the tether, avionics, payload ejection and payload support systems including data systems, end mass attitude control, communication and data collection.

For further information please refer to the *AIRSEDS Internet Central* web site at http://www.airseds.com/.

Contacts:
• A. Santangelo - The Michigan Technic Corporation

2.5 The RAPUNZEL Mission

The small tether project RAPUNZEL was started in 1991 by the Institute of Astronautics, Munich Technische Universitat (TU) and the Kayser-Threde Company to design a low cost tether experiment. In collaboration with the Samara State Aerospace University (SSAU), Russia, the initial mission intended to fly the German re-entry capsule MIRKA on a Russian Photon capsule. Later on, in collaboration with SSAU and the former NPO Energia, the project split into three different missions on Resurs, Photon, and Progress spacecraft, respectively.

The TU team designed and built a deployer based on textile technology, which would ensure both high reliability and low cost (Fig. 1). SSAU is building a small re-entry capsule to fly on Resurs.

Lately, the main effort has been the development and test of the deployer. In November 1995, a campaign of parabolic flights tested the deployer under microgravity conditions. The first tests have shown good results and proven the concept feasibility. The laboratory tests were followed by numerical simulations of the payload deployment and its re-entry in the atmosphere.

Fig. 2 shows a schematic of the mission sequence. When the endmass is ejected by springs (1) the deployment starts (2). After reaching the full tether length of 52 km, the tether is cut (3) and the capsule reenters the earth atmosphere (4) and lands on parachute (5). Preliminary simulations have shown that even though the atmospheric drag induces small oscillations in the system, the endmass lands safely in the Kasakstan region.

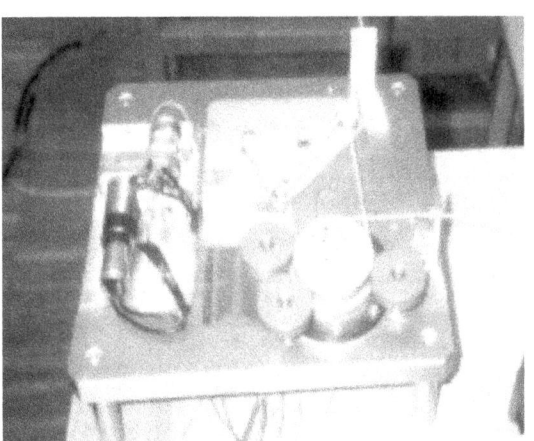

Figure 1. Breadboard model of tether deployer

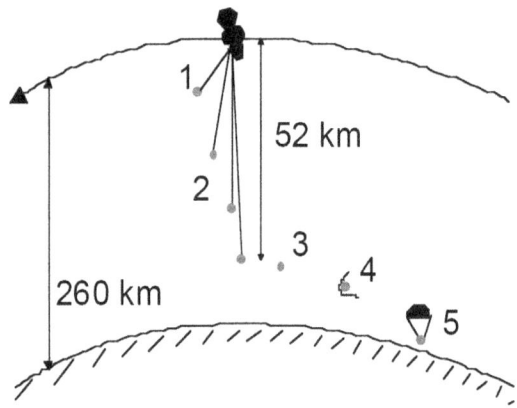

Figure 2. Schematic view of the deployment sequence

Contacts:

- Manfred Krischke, Kayser-Threde GmbH, Munich, Germany
- Dieter Sabath - Technische Universitat Munich, Germany

2.6 Tether Mechanism Materials and Manufacture Project

The ESA funded Tether Mechanism Materials and Manufacture (TMM&M) project has been performed by Alenia Spazio (Italy), as prime contractor, and SABCA (Belgium) and SENER (Spain) as subcontractors.

One important class of low-cost tether mechanisms and related space missions was identified in the development of expendable tether systems that did not require complex mechanism operations and the associated technology development. For the TMM&M ESA technology development activity, a EURECA-based tether initiated material or sample re-entry model mission, with a 150-kg mass capsule and a 20-km tether, was adopted for the expendable tether mechanism design and its breadboard model selected to be manufactured and tested.

A particular challenge in the expendable tether mechanism design, associated with a near-horizontal tether deployment operation, was represented by the deployment control, tension and rate ranges and accuracy requirements. Various simple tether mechanism design solutions were traded-off and a spool-reel configuration solution, in which no (passive) control is applied in the early tether spool deployment operation and active reel-brake (rate-feedback) actions are implemented to control the remaining part of deployment accurately, was adopted and bread-boarded. The TMM&M Project expendable tether mechanism bread-board model (fig. 1) was functionally tested on a suitably designed and manufactured test facility capable of performing tether deployment testing for a vast range of preselected length, rate and tension reference profiles.

Figure 1. Expendable tether mechanism breadboard model

Contacts:
- R. Licata, P. Merlina - Alenia
- J.M. Gavira - ESA/Estec

2.7 The Space Tether Experiment (STEX)

Description

The Space Tether Experiment (STEX) has been proposed by ISAS to fly onboard the Space Flight Unit (SFU) follow-on mission as one of the science and technology experiments.

SFU orbit is 500 km and circular and the spacecraft attitude is sun-oriented. SFU can carry 1000 kg of payload. The major objective of STEX is to assess the tether technology for future scientific missions.

Mission Scenario

In order to evaluate the performance of different control logics, a 40-kg subsatellite will be deployed up to 10 km and retrieved several times. During stationkeeping the susbatellite will be stabilzed along the vertical with impulsive thrusts.

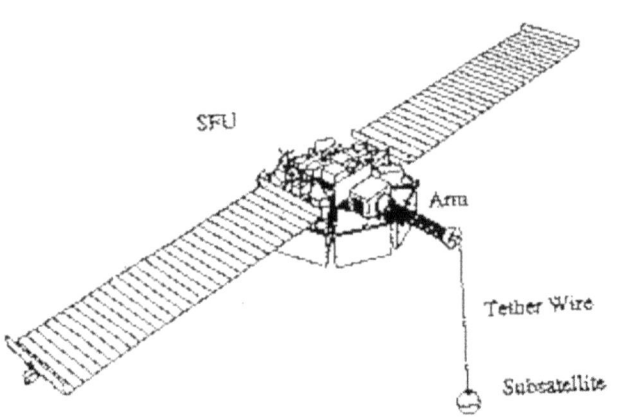

Figure 1. STEX on board SFU

Instrumentation

The subsatellite will be equipped with a vacuum gauge, plasma probes and wave receivers to study SFU electromagnetic environment. A tether deployment and retraction system has been developed for laboratory tests, a schematic is shown in figure 2. The deployment/retrieval speed, tether tension and reponse of the feedback system have been analyzed using this system.

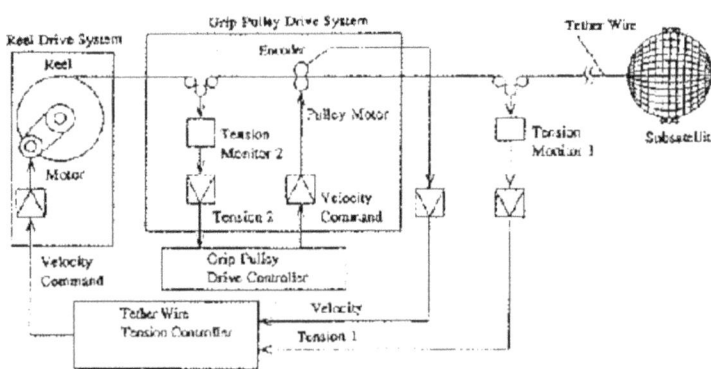

Figure 2. Schematic of STEX Deployment/Retrieval System

Contacts:
• K.I. Oyama, S.Sasaki, ISAS

54

SECTION 3.0 TETHER APPLICATIONS

3.1 General

This section provides a summary of various tether applications proposed thus far, concentrating on near-term, mid-term, and innovative applications. In some cases, these applications are general ideas, and in others, they are well-defined systems, based on detailed study and computational analysis. These applications have been divided into eight general categories. In cases where an application can be logically placed in more than one, it has been placed in the one considered most appropriate. To avoid redundancy, variations of a particular system concept are not described separately. Instead, Section 3.2 contains a listing of the applications by category, page number, and possible cross reference to other categories. Descriptions of proposed applications follow this listing. For these descriptions, a standardized format is used to allow quick and easy comparisons of different applications. This format is designed to effectively serve as wide a readership as possible, and to conveniently convey the pertinent details of each application. Readers with different interests and needs can find the information and level of detail they desire at a glance.

The Category and title of each application is presented at the top of the page. The "Application" subsection provides a brief statement of the application, and the "Description" subsection provides a brief description of the system design and operation. A picture is located in the upper right of the page to supplement the description, by providing a diagrammatic representation of the system and its operation. The "Characteristics" subsection exhibits the major system design and operation parameters in bullet form. The last characteristic is always a bullet entitled "Potential for Technology Demonstration". This entry attempts to classify both the conceptual maturity of an application, and the amount of technological development required to demonstrate the particular application. When applicable we have mentioned the availability of flight data that somehow may support the feasibility of the application. Three descriptors have been used to indicate the demonstration time-frame:

- Near-Term: 5 years or less,
- Mid-Term: 5-10 years, and
- Far-Term: 10 years or greater.

The date of this printing may be assumed to be the beginning of the Near-Term period. Together, these subsections present a brief and complete summary of the system's application, design, and operation.

The "Critical Issues" subsection, lists the developmental and operational questions and issues of critical importance to the application. The "Status" subsection indicates the status of studies, designs, development, and demonstrations related to the application. The "Discussion" subsection presents more detailed information about all aspects of the application. Following this, the "Contacts" subsection lists the names of investigators who are involved with work related to the application, and who may be contacted for further information. (See "Contacts" section, for addresses and telephone numbers.) Finally, the "References" subsection lists the reference used in the preparation of the application description.

Many of the applications that follow are subject to similar critical issues which are more or less "generic" to tethers. These are issues such as damage from micrometeoroids or other space debris, dynamic noise induced on platforms, high power control electronics technology, rendezvous guidance and control, tether material technology development, and system integration. Many of the figures presented in the "Tether Data" section address these critical issues.

3.2 Tether Applications Listing

Following is a list of abbreviations used to identify cross references to other categories. The application listing has been arranged in alphabetical order by category and application within each category.

AE   AERODYNAMICS
CN   CONCEPTS
CG   CONTROLLED GRAVITY
EL   ELECTRODYNAMICS

PL   PLANETARY
SC   SCIENCE
SS   SPACE STATION
TR   TRANSPORTATION

3.3 Tether Applications

-- AERODYNAMICS --

Station Tethered Express Payload System (STEPS)

APPLICATION: Provides a way to return small payloads from the International Space Station to earth between shuttle flights, without the safety hazards of handling rocket motors or propellants.

DESCRIPTION: Payloads are tied down inside a mini-Apollo capsule small enough to fit through the robotic airlock in the Japanese Experiment Module. The capsule is ejected downward and deploys using a SEDS-1 (deploy-swing) strategy. The tether is cut free at the station end, and it orients the capsule for reentry before burning off. (This "kite tail" effect was validated by SEDS-1.)

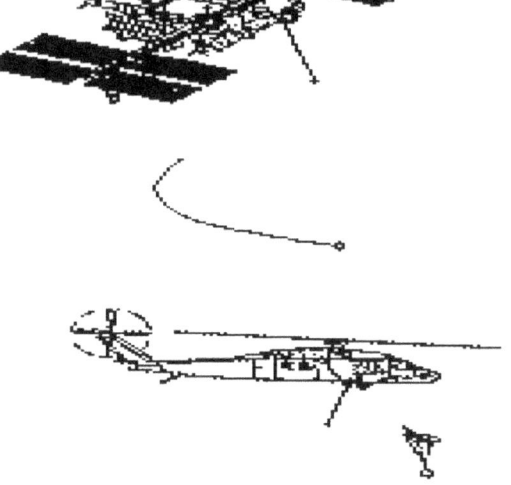

CHARACTERISTICS:
- Tether length: 30-33 km
- Payload: 30 kg, 100 liters
- Timescale: Near-Term

CRITICAL ISSUES:
- Tether deployment control for proper swing
- Implications of micrometeoroid cut (~0.7% risk)
- Accelerations of ~4 microgee on station during swing

STATUS:
- Tether Applications has contract to deliver protoflight capsule & deployer Feb 1998.
- Capsule can be tested as Delta or Progress secondary payload; both are under study.

DISCUSSION: The tether deployer is a smaller easily reloadable version of SEDS. It mounts in a reusable capsule balancer/ejector/deployer assembly that remains with the station.
For test flights, the deployer and flight computer mount inside the capsule. This simplifies integration on the host vehicle and maximizes hardware recovery for inspection and potential re-use. Baseline recovery scenario involves soft mid-air capture of gliding parachute by helicopter.

CONTACTS:
- Joe Carroll
- Chris Rupp
- Paul Kolodziej

REFERENCES:
>   A Station Tethered Express Payload System (STEPS), available from Tether Applications

-- AERODYNAMICS --

Multiprobe for Atmospheric Studies

APPLICATION: Measurement of spatial geophysical gradients.

DESCRIPTION: A one-dimensional
constella-tion of probes is lowered by the Shuttle
or Space Station into the atmosphere in order to
provide simultaneous data collection at different
locations.

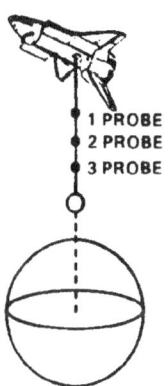

CHARACTERISTICS:
- •   Physical
       Characteristics:Mission related
- •   Potential For
       Technology
       Demonstration: Near-Term

CRITICAL ISSUES:
- •   Crawling systems might be necessary
- •   Operational sequence for deployment and retrieval

STATUS:
- •   Configuration study performed by Smithsonian Astrophysical Observatory
- •   Analysis of scientific applications performed at University of Texas ,
       Dallas

DISCUSSION: This constellation configuration could prove very valuable in low altitude
measurements requiring simultaneous data collection at the various probe positions. Good
time correlation of the measurements is one benefit of this system.

CONTACTS:
- •   Enrico Lorenzini
- •   Rod Heelis

REFERENCES:
       Proc. of Fourth International Conference on Tethers in Space, Washington DC,
       10-14 April 1995

-- AERODYNAMICS --

Shuttle Continuous Open Wind Tunnel

APPLICATION: Obtain steady-
state aerothermo-dynamic research data
under real gas conditions without
experiencing limiting effects inherent
in ground-based wind
tunnels.

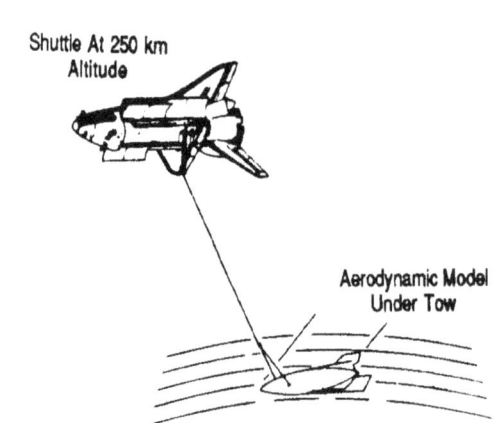

Shuttle At 250 km
Altitude

Aerodynamic Model
Under Tow

DESCRIPTION: A tethered
aerodynamically shaped research vehicle
is deployed downward form the Space
Shuttle to obtain data in the free
molecule, transition, and upper
continuum flow regimes.
Characterization of the free-stream,
measurement of gas-surface interactions,
flow field profiling, and determination of
state vectors are to be accomplished.

CHARACTERISTICS:
- •   Length: 100-120 km
- •   Mass: Variable, dependent on mission requirements
- •   Power Required: TBD, for instruments and data handling only
- •   Potential For
         Technology
         Demonstration: Near-Term

CRITICAL ISSUES:
- •   Quantitative definition of data requirements
- •   Define method for flow-field profiling
- •   Quantitative analysis of orifice effects vs. altitude

STATUS:
- •   Prototype experiment and instrument package proposed for ATM mission

DISCUSSION: Unique measurements are possible due to low Reynold;s number and high
Mach number regime. Measurements in real-gas will provide more dependable data
regarding fluid flow, turbulence, and gas-surface interactions.

CONTACTS:
- •   Giovanni Carlomagno
- •   Franck Hurlbut
- •   George Wood

REFERENCES:
Proc. of Fourth International Conference on Tethers in Space, Washington DC,
10-14 April 1995

-- CONCEPTS --

Gravity Wave Detection Using Tethers

APPLICATION: To detect gravity
waves from sources such as binary stars,
pulsars, and supernovae.

DESCRIPTION: The system would consist of two
masses on each end of a long tether with a spring at its
center. As this tether system orbits the Earth,
gravitational waves would cause the masses to oscillate.
This motion would be transmitted to the spring, which
would be monitored by a sensing device. Analysis of the
spring displacement and frequency could then lead to the
detection of gravity waves.

CHARACTERISTICS:
- Mass: 20 kg (Each End Mass)
- Tether Length: 25 km
- Tether diameter: 0.6 mm
- Spring Constant: $K_S = 2.3 \times 10^3$ dyne/cm
- Orbital Altitude: ≥ 1000 km   • Potential For Technology Demonstration: Long-Term

CRITICAL ISSUES:
- Existence of gravity waves
- Gravity wave noise level from other bodies
- Excitation of oscillations from other sources

STATUS:
- Preliminary calculations have been performed at SAO, Caltech, and Moscow State University

DISCUSSION: This gravitational wave detector would operate in the 10 - 100 MHz
frequency band that is inaccessible to Earth-based detectors because of seismic noise. If
gravitational waves do exist in this region, a simple system such as a tether-spring detector
would prove of great value.

CONTACTS:
- K. Thorne
- Marino Dobrowolny

REFERENCES:
V.B. Braginski and K.S. Thorne, "Skyhook Gravitational Wave Detector," Moscow
State University, Moscow, USSR, and Caltech, 1985.

B. Bertotti, R. Catenacci, M. Dobrowolny, "Resonant Detection of Gravitational Waves by Means of Long Tethers in Space," Technical Note (Progress Report), Smithsonian Astrophysical Observatory, Cambridge, Massachusetts, March 1977.

-- CONCEPTS --

Tethered Lifting Probe

APPLICATION: The lifting body controls the altitude of the probe in atmospheric tether missions.

DESCRIPTION: A hypersonic lifting body is used for the probe in an atmospheric mission. Changes in lift forces on the probe can be used to control the probe altitude without changing the length of the tether. Required changes in probe attitude can be accomplished using a movable tether attachment point or aerodynamic control surfaces.

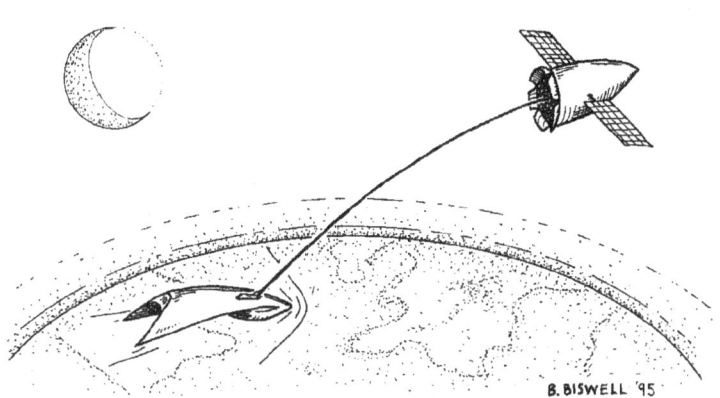

CHARACTERISTICS:
- Tether Length: 10-200 km
- Probe Area: 10-50 m^2
- Potential For Technology Demonstration: Mid-Term

CRITICAL ISSUES:
- Development of control laws to maintain probe attitude.

STATUS:
- Preliminary results indicate the feasibility of using lift as a control mechanism for probe altitude.
- Current studies favor the use of a movable tether attachment point as a simple and highly effective attitude control mechanism.

DISCUSSION: The lifting probe provides an ideal control mechanism for the altitude of an atmospheric tether system. The alternative is to slowly change the tether length by using a reel mechanism. This may not be effective in situations where probe altitude must be maintained in the presence of atmospheric uncertainties. In addition, the use of a lifting body can increase the atmospheric penetration of the probe without increasing its mass. This concept can be applied to a wide range of tether atmospheric missions from upper atmosphere research to aerocapture.

CONTACTS:
- Jordi Puig-Suari
- Brian Biswell

REFERENCES:
Biswell, B., and Puig-Suari, J. "Lifting Body Effects on the Equilibrium Orientation of Tethers in the Atmosphere,"AIAA-96-3597, *AIAA/AAS Astrodynamics Conference*, San Diego, CA, 1996.
Keshmiri, M., and Misra, A.K. "Effects of Aerodynamic Lift on the Stability of Tethered Subsatellite System," AAS-93-184, *AAS/AIAA Spaceflight Mechanics Meeting*, Pasadena, CA, 1993.

External Tank Space Structures

APPLICATION: Utilize Shuttle external tanks in a raft format to form a structure in space.

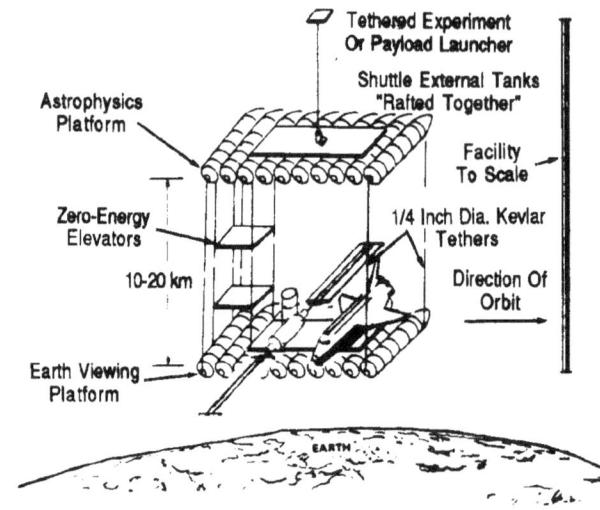

DESCRIPTION: Tethers are used to separate rafts composed of external tanks. These can either be used as a "Space Station" or as structural elements in an evolving Space Station.

CHARACTERISTICS:
- Tether Length: 10 - 20 km
- Potential For Technology Demonstration: Long-Term

CRITICAL ISSUES:
- Space operations required to adapt tanks to proposed applications
- External tank induced contamination environment
- Stability/controllability of proposed configuration
- Assembly/buildup operations
- Drag makeup requirements

STATUS:
- Preliminary analysis performed
- Further analyses effort deferred

DISCUSSION: Most likely use of this concept would be as a "space anchor" for tether deployment concepts.

CONTACTS:
- Joe Carroll

REFERENCES:
Carroll, J. A., "Tethers and External Tanks, Chapter 3 of Utilization of the External Tanks of the Space Transportation System," California Space Institute, La Jolla, California, Sept. 1982.

Carroll, J. A., "Tethers and External Tanks: Enhancing the capabilities of the Space Transportation System," Dec. 1982

Heliocentric Alfven Engine for Interplanetary Transportation

APPLICATION: Generation of propulsion for interplanetary travel by using the electromagnetic interaction of a conducting tether and the interplanetary magnetic field.

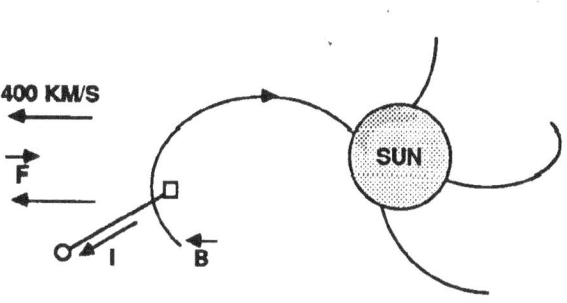

DESCRIPTION: An insulated conducting tether, connected to a spacecraft and terminated at both ends by plasma contactors, provides interplanetary propulsion in two ways. The current induced in the tether by the solar wind magnetic field is used to power ion thrusters. The interaction between the tether current and the magnetic field can also be used to produce thrust or drag.

CHARACTERISTICS:
- Tether Length: 1000 km
- Cooling: Helium (2°K)
- Current: 1000 A
- Power: 2 MW
- Materials: Superconducting Niobium-Tin
- Potential For Technology Demonstration: Far-Term

CRITICAL ISSUES:
- How does this system compare with others, such as nuclear or solar sail
- Feasibility and controllability have not been established

STATUS:
- TSS-1R flight to demonstrate electrodynamic interaction with surrounding plasma
- More detailed study and evaluation of this application are required

DISCUSSION: The solar wind is a magnetized plasma that spirals outward from the sun with a radial velocity of about 400 km/sec. The magnetic field of the solar wind is 5×10^{-5} Gauss, producing an electric field of 2 V/km, as seen by an interplanetary spacecraft. If a conducting tether, connected to the spacecraft and terminated at both ends by plasma contactors, were aligned with the electric field, the emf induced in it could yield an electric current. This current could be used to power ion thrusters for propulsion. The current could be maximized by using superconducting materials for the tether. (This system was proposed by Hannes Alfven in 1972). It has been calculated that a 1000 km superconducting wire of Niobium-tin could generate 1000 A (2 MW). To achieve superconduction temperatures, this wire could be housed in an aluminum tube with flowing supercooled (2° K) helium. The tube would be insulated and capped at each end with a refrigeration system.

In addition to the ion thrusters, the interaction of the tether current and solar wind magnetic field would produce thrust or drag. As current flowed in the tether, the magnetic field would exert an IL x B force on the tether. If the spacecraft were moving away from the sun (with the solar wind), a propulsive force would be exerted on the tether as its electrical power was dissipated. A drag would be exerted on the tether if current from an on-board power supply were fed into it against the induced emf. When moving toward the sun (against the solar wind), the opposite conditions would apply.

This system could be used to spiral away from or toward the sun, or to move out of the ecliptic. Theoretically, such a spacecraft could attain the solar wind velocity of 400 km/sec. Use of the electromagnetic interaction between a conducting tether system and the solar wind may allow much shorter transfer times and larger payloads for planetary missions.

CONTACTS:
- Mario Grossi
- Jim McCoy
- Nobie Stone

REFERENCES:
Applications of Tethers in Space, Vol. 1,2 Workshop Proceedings, NASA CP-2365, March 1985

H. Alfven, "Spacecraft Propulsion: New Methods," Science, Vol. 176, pp. 167-168, April 14, 1972.

-- CONCEPTS --

Earth-Moon Tether Transport System

APPLICATION:
Transportation of material from
lunar to Earth orbit.

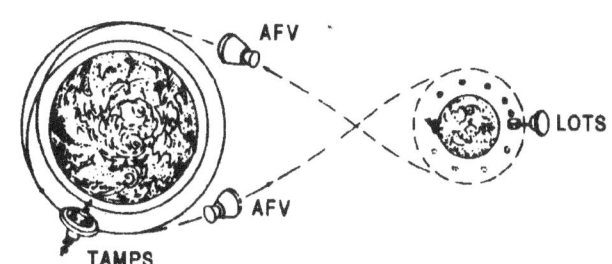

DESCRIPTION: Material
(probably Moon rocks) in lunar
orbit is collected by the LOTS
(Lunar Orbiting Tether Station),
half is transferred to an AFV
(Aerobraking Ferry Vehicle)
which transports it to LEO,
where it is transferred to the TAMPS (Tether And Materials Processing Station). The AFV
then returns to the Moon for more lunar material.

CHARACTERISTICS:
- Physical Characteristics: Undetermined
- Potential For   Technology Demonstration: Far-Term

CRITICAL ISSUES:
- Undetermined

STATUS:
- No detailed study on this application has been performed

DISCUSSION: Material (probably Moon rocks) in lunar orbit could be transported to Earth
orbit without the use of propellants with this tether transport system. (The material in
lunar orbit could have been placed there by the Lunar Equator Surface Sling; Application
"Lunar Equator Surface Sling"). It could be collected in orbit by a Lunar Orbiting Tether
Station (LOTS). The LOTS would proceed as follows: (1) catch the rocks, spin-up, catch
an Aerobraking Ferry Vehicle (AFV); (2) Load the AFV with half of the rocks; (3) spin-up,
throw the AFV into trans-Earth injection; (4) de-spin, load the other rocks on a tether; and
(5) spin-up and deboost the rocks for momentum recovery.
The AFV would proceed to Earth, where it would aerobrake into LEO for capture by the
Tether And Materials Processing Station (TAMPS). The TAMPS would proceed as
follows: (1) catch, retrieve, and unload the aerobraked AFV; (2) process moonrocks into
LO_2, etc; (3) refuel and reboost the AFV toward the Moon; (4) recover momentum with an
electromagnetic tether; and (5) also capture, refuel, and reboost AFV's going to GEO and
deep space when required. The AFV returning to the Moon would be a rocket boosted into
trans-lunar injection and final lunar orbit for recapture by the LOTS.

CONTACTS:
- Joe Carroll

REFERENCES:
Applications of Tethers in Space, NASA CP-2422, March 1986.

69

-- CONCEPTS --

Mars Moons Tether Transport System

APPLICATION: Transportation of manned vehicles and spacecraft from low Mars orbit out to escape, or from escape to low Mars orbit, using tethers attached to the Moons of Mars.

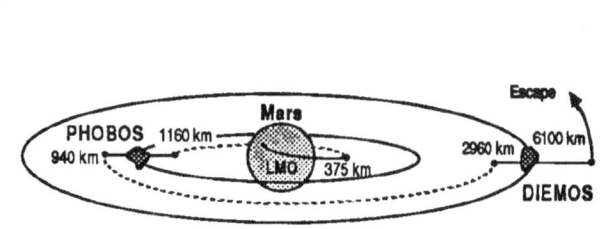

DESCRIPTION: Long tethers (Kevlar strength or better) are attached above and below both Phobos and Deimos to ferry vehicles and other payloads between low Mars orbit and Mars escape without the use of propulsion. For example, a vehicle is tethered upward from a low Mars orbit station, released, and then caught by a downward hanging tether on Phobos. The payload is then transferred to the upward deployed tether and released. The process is repeated at Deimos, and results in escape from Mars. The process is reversible.

CHARACTERISTICS:
- Length: 940 km (up), 1160 km (down) at Phobos
                             6100 km (up), 2960 km (down) at Deimos
- Tether Mass: 5000 kg to 90,000 kg
- Tether Diameter: 2 mm (or greater)
- Power: TBD
- Materials: Kevlar, or higher strength material
- Payload Mass: 20,000 kg
- Potential For Technology Demonstration: Far-Term

CRITICAL ISSUES:
- Tether dynamics analysis
- Comparison with other advanced propulsion methods
- Rendezvous feasibility
- Operations and cost
- Tether severing by micrometeoroids or debris

STATUS:
- A conceptual study defines the tether length and strength requirements, but does not address construction, placement, and operation of the tether station.

DISCUSSION: The two moons of Mars, Phobos and Deimos are near equatorial, and can function as momentum banks in the transfer of mass from Mars low orbit to Mars escape (or the reverse). The requirement is to place long tethers, upward and downward, on each of the two moons of Mars. Example uses might be to transfer Deimos or comet material

70

to the Mars surface or to transfer astronauts from Mars surface to a waiting interplanetary low thrust vehicle at Deimos, or to support materials processing in Mars orbit.

Tether stations on Phobos and Deimos may have to be manned for construction, operation, and maintenance. Therefore, other human functions at these satellites would be necessary to make this concept viable. It is best suited to a high activity scenario with departures and arrivals at Mars daily or weekly. A station on Phobos alone would be sufficient for near Mars operations, and could even be used for escape with a sufficiently long upward tether. The mass of the two bodies is so great, ($>10^{15}$ kg) that their orbits would not be affected for decades or longer.

CONTACTS:
- Joe Carroll
- Paul Penzo

REFERENCES:

Penzo, P. A., "Tethers for Mars Space Operations," The Case for Mars II, Ed. C. P. McKay, Vol. 62, Science and Technology Series, p. 445-465, July 1984.

-- CONTROLLED GRAVITY --

Rotating Controlled-Gravity Laboratory (Tethered Platform)

APPLICATION: Provide a readily accessible variable/controlled gravity laboratory, capable of generating artificial gravity levels of up to 1 g and over, in Earth orbit.

DESCRIPTION: A tethered platform composed of two end structures, connected by a deployable/retractable 10 km tether. One end structure includes the solar arrays, related subsystems, and tether reel mechanism. The other includes two manned modules and a propellant motor. Artificial gravity is created in the manned modules by extending the tether and firing the motor, rotating the entire system about its center of mass (the solar panels are de-spun). Tether length is used to control the gravity level.

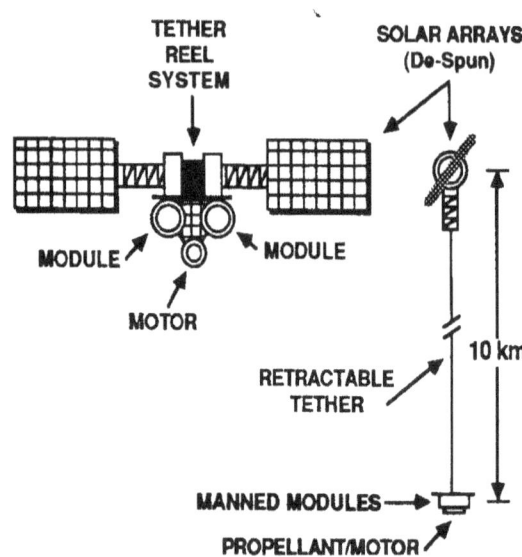

TETHER PLATFORM CONCEPT

CHARACTERISTICS:
- Length: Up to 10 km
- g-Level: Up to 1.25
- Rotation Rate: Up to 0.75 rpm
- Potential for Technology Demonstration: Far-Term

CRITICAL ISSUES:
- Susceptibility to micrometeoroid/debris damage

STATUS:
- A detailed dynamic analysis has been performed at SAO
- A System study has been performed at Stanford University

DISCUSSION: Access to an orbiting variable/controlled-gravity laboratory, capable of providing artificial gravity levels of up to 1 g and over, would allow vital experimentation in this important gravity range, and provide an appropriate facility, should artificial gravity be determined to be a physiological requirement for extended manned orbital missions. Artificial gravity (in the form of centrifugal acceleration) would be created by rotating the laboratory. The magnitude of the resulting centrifugal acceleration is equal to the square of the angular velocity times the radius of rotation.

Three basic rotating lab configurations are possible - a torus or cylinder (centrifuge), a rigid station, and a tethered platform. The centrifuge is the least attractive because of its relatively small volume, large Coriolis force, and large dynamic disturbance levels. Of the remaining two, the tethered system has several advantages over the rigid one. It would provide a larger radius of rotation, reducing the rotational rate required to produce a desired g-level. This, in turn, would reduce unwanted side effects, such as the Coriolis force. The variable tether length would also allow a large variety of artificial gravity environments. To spin the system, the tether would be extended to its full 10 km

72

length, and the motor fired. (The minimum necessary Delta-V has been calculated to be 125 m/s.) The tether length would then be adjusted to provide the desired g-level. Assuming the end masses are equal and rotating about a common center, 0.08 g would result from a tether length of 10 km at a spin rate of 0.12 rpm, 0.16 g (lunar gravity) from a length of 8 km at 0.20 rpm, 0.38 g (Mars gravity) from a length of 6 km at 0.33 rpm, 1 g from a length of 4.3 km at 0.65 rpm, and 1.25 g from a length of 4 km at 0.75 rpm. The solar arrays would be de-spun and sun-oriented. However, a disadvantage is the high Delta-V required to start and stop this spin. Another is the fact that the rotation would probably have to be stopped to allow docking with a spacecraft.

This lab would allow experimentation at gravity levels ranging from low gravity, through Moon, Mars, and Earth gravities, to more than 1 g. The effects of gravity on plant and animal growth, and on human performance and medical processes (such as those related to the cardiovascular, skeletal, and vestibular systems) could be studied for prolonged periods of time. Gravity conditions on the Moon and Mars could be simulated, and the lab could be used to prepare for the possible use of artificial gravity on manned interplanetary missions. It could also provide Earth-like habitability at partial g. Such physical processes as crystal growth, fluid science, and chemical reactions could be studied at various gravity levels.

CONTACTS:
- Enrico Lorenzini
- Paul Penzo
- Chris Rupp

REFERENCES:
Applications of Tethers in Space, NASA CP-2422, March 1986

B.M.Quadrelli, E.C. Lorenzini, "Dynamics and Stability of a Tethered Centrifuge in Low Earth Orbit", The Journal of the Astronautical Sciences, Vol. 40, No. 1, 1992, pp.3-25

Powell, J. David, Systems Study of a Variable Gravity Research Facility, Final Report to NASA (Grant No. NCA2-208), April 1988.

-- CONTROLLED GRAVITY --

Tethered Space Elevator

APPLICATION: The Space Elevator may
be used as a Space Station facility to tap different
levels of residual gravity, and a transportation
facility to easily access tethered platforms.

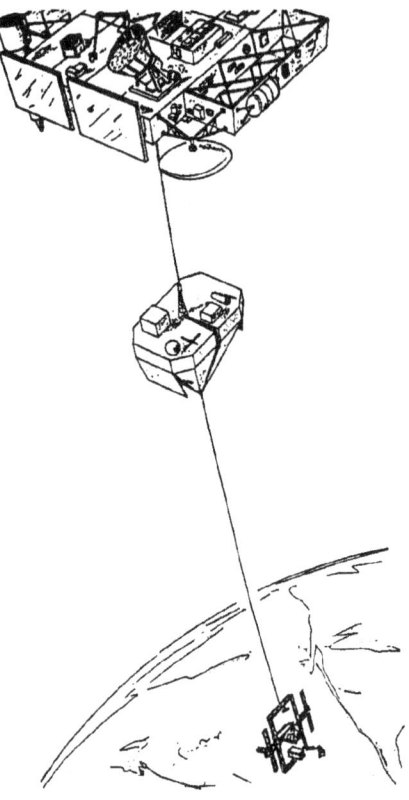

DESCRIPTION: The Space Elevator is an
element able to move along the tether in a
controlled way by means of a suitable drive
mechanism. The primary objectives of the
microgravity elevator mission are the
achievement of a new controllable microgravity
environment and the full utilization of the Space
Station support while avoiding the microgravity
disturbances on board the Space Station. A
shorter and slack cable could be used as both a
power and data link.

A ballast mass represents the terminal end
of the tether system. It could be any mass (e.g., a
Shuttle ET) or a tethered platform. The
objective of the transportation elevator
application is to access large tethered platforms
for maintenance, supply of consumables, or
module and experiment exchanges.

CHARACTERISTICS:

- Length: 10 km
- Elevator Mass: 5,000 kg
- Ballast Mass: Up to 50,000 kg
- g-Level: 10^{-7} to 10^{-3}
- Power Required: Up to 10 kW by Tether   •   Potential For
                       Power Line Link   Technology
- Link Data Rate: Up to 40 Mb/s by Tether   Demonstration: Mid-Term
                       Optical Fiber Link

CRITICAL ISSUES:

- Space Station impacts
- Dynamic noise induced on the tether drive mechanism
- Gravity-measuring instrumentation
- Power link technology
- Optical fibers link technology

STATUS:

- ASI/Aeritalia Elevator Definition Study in initial design assessment phase,
  Final Report issued in March 1988

- Analysis of dynamics during deployment, station-keeping, and transfer maneuvers carried out by the Smithsonian Astrophysical Observatory under contract to NASA/MSFC

DISCUSSION: The most promising feature offered by the Space Elevator is the unique capability to control with time the gravity acceleration level. In fact, since the radial acceleration changes with position along the tether, the Elevator would be able to attain a continuous range and a desired profile vs. time of residual gravity level by the control of the Elevator motion. Moreover, the Elevator is able to fully utilize the Space Station support (power, communications, logistics) and to avoid the Space Station contaminated environment, from a microgravity point of view, by tether mediation.
Another way to exploit the Space Elevator capabilities is its utilization as a transportation facility. The idea of using large tethered platforms connected to the Space Station by power line and communication link (via tether technology) makes unrealistic frequent operations of deployment and retrieval. On the other hand, the platform may require easy access for maintenance, supply of consumables, module and experiment exchange. The Space Elevator, as a transportation facility able to move along the tether to and from the platform, may be the key to tethered platform evolution.

CONTACTS:
- Franco Bevilacqua
- Enrico Lorenzini
- Pietro Merlina

REFERENCES:

Applications of Tethers in Space, NASA CP-2422, March 1986

F. Bevilacqua and P. Merlina, "The Tethered Space Elevator System," Second International Conference on Tethers In Space, Venice, Italy, 1987.

SATP Definition Study, Mid-Term Report, Aeritalia, TA-RP-AI-002, March 21, 1986.

Tethered Space Elevator Definition and Preliminary Design, Final Report, Aeritalia, TA-RP-AI-009, 1988.

L.G. Napolitano and F. Bevilacqua, "Tethered Constellations, Their Utilization as Microgravity Platforms and Relevant Features," IAF-84-439.

S. Bergamaschi, P. Merlina, "The Tethered Platform: A Tool for Space Science and Application," AIAA-86-0400, AIAA 24th Aerospace Sciences Meeting, Reno, Nevada, January 6-9, 1986.

Lorenzini, E.C., M.D. Grossi, D.A. Arnold, and G.E. Gullahorn, "Analytical Investigation of the Dynamics of Tethered Constellations in Earth Orbit (Phase II)," Smithsonian Astrophysical Observatory Reports for NASA/MSFC, Contract NAS8-36606. Quarterly Reports

Lorenzini, E.C., "A Three-Mass Tethered System for Micro-g/Variable-g Applications," Journal of Guidance, Control, and Dynamics, Vol. 10, No.3, May-June 1987. (pp. 242-249)

Applications "Microgravity Laboratory" and "Variable/Low Gravity Laboratory"

75

-- ELECTRODYNAMICS --

Electrodynamic Power Generation (Electrodynamic Brake)

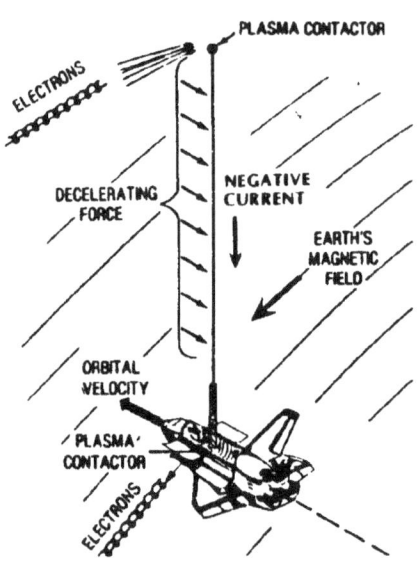

APPLICATION: Generation of DC electrical power to supply primary power to on-board loads.

DESCRIPTION: An insulated conducting tether connected to a spacecraft and possibly terminated with a subsatellite. Plasma contactors are used at both tether ends or with the bare tether (see sect. 2). Motion through the geomagnetic field induces a voltage across the orbiting tether. DC electrical power is generated at the expense of spacecraft/tether orbital energy.

CHARACTERISTICS:
- Power Produced: 1 kW - 1 MW
- Length: 10 - 20 km
- Mass: 900 - 19,000 kg
- Efficiency: ~90%
- Materials: Aluminum
- Potential For Technology Demonstration: Near-Term

CRITICAL ISSUES:
- Flight experiment validation of the current-voltage characteristics of plasma contactor devices and operating at currents of up to 50 A in the ionosphere are urgently needed to validate results from chamber tests and theoretical models in space
- Flight experiment validation of the current-voltage characteristics of the bare tether concept
- Flight experiment determination of the role played by ignited mode operation in the ionosphere
- Ground and flight experiment validation of the theoretically predicted role of plasma contactor cloud instabilities
- Characterization of the magnetosphere current closure path and its losses
- Characterization of the effects of large electromagnetic tether systems on the LEO environment and other space vehicles
- Assurance of long-term insulator life
- Characterization of massive tether dynamics
- Development of space compatible insulation methods and power processing electronics for multikilovolt operation
- Susceptibility to micrometeoroid/debris damage
- Understanding of current collection effects at resulting insulator defects and their impacts on system performance (as in TSS1R)

- TSS-1 and -1R, PMG flights
- A wide variety of work is actively underway in the areas of electrodynamic demonstrations, hollow cathodes, tether materials, and hardware technologies including a demo flight (see section 2 and "bare tether" concept)

DISCUSSION: An orbiting insulated tether, terminated at the ends either by plasma contactors or by a bare section of tether, can be used reversibly as an electrical power or thrust generator. Motion through the geomagnetic field induces a voltage in the tether, proportional to its length and derived from the v x B electric field and its force on charges in the tether. This voltage can be used to derive a DC electrical current in the tether. Electrical power is generated at a rate equal to the loss in spacecraft orbital energy due to a drag force of magnitude (ilB) where i is the tether current and l is the length. It has been shown that this drag force functions as an electrodynamic brake and can be used to perform orbit maneuvering in LEO or in the ionosphere of planets such as Jupiter or Saturn.

Three basic plasma contactor configurations have been considered in the studies performed to date: (1) a passive large-area conductor at both tether ends; (2) a passive large-area conductor at the upper (positive) end and an electron gun at the lower (negative) end and (3) a plasma-generating hollow cathode configuration. Hollow cathodes as flown on PMG are considered to be safer for spacecraft systems, since they establish a known vehicle ground reference potential with respect to the local plasma. They also allow simple reversibility of the tether current for switching between power and thrust generation.

CONTACTS:
- Les Johnson
- Joseph Kolecki
- Jim McCoy
- Juan Sanmartin
- Nobie Stone

REFERENCES:

Proc. of Fourth International Conference on Tethers in Space, Washington DC, 10-14 April 1995

Electrodynamic Thrust Generation

APPLICATION: Generation of electro-magnetic propulsive thrust to boost the orbit of a spacecraft.

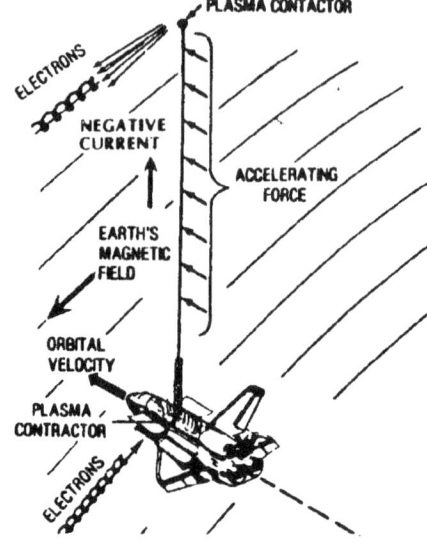

DESCRIPTION: An insulated conducting tether connected to a spacecraft and possibly terminated with a subsatellite. Plasma contactors are used at both tether ends. Current from an on-board power supply is fed into the tether against the emf induced by the geomagnetic field, producing a propulsive force on the spacecraft/tether system. The propulsive force is generated at the expense of primary on-board electric power.

CHARACTERISTICS:
- Thrust Produced: Up to 200 N
- Power Required: Up to 1.6 MW
- Length: 10-20 km
- Mass: 100-20,000 kg & power supply
- Efficiency: ~90%
- Materials: Aluminum
- Potential For Technology Demonstration: Near-Term

CRITICAL ISSUES:
- The same as listed in Electrodynamic Power Generation application

STATUS:
- The same as listed in Electrodynamic Power Generation application

DISCUSSION: An insulated conducting tether, terminated at the ends by plasma contactors, can be used reversibly as an electromagnetic thruster or electrical power generator. A propulsive force of IL x B is generated on the spacecraft/tether system when current from an on-board power supply is fed into the tether against the emf induced in it by the geomagnetic field.

Recommendations have been made through the years to use electrodynamic tethers to provide drag compensation and orbital maneuvering capability for the International Space Station, other solar array powered satellites, and to use higher power tethers (up to about 1 MW) for orbital maneuvering of the Space Station and other large space systems. Design tradeoffs were also recommended, including:

- Use of counterbalancing tethers deployed in opposite directions to provide center-of-mass-location control
- Use of shorter tethers operating at low voltage and high current versus longer tethers operating at high voltage and low current
- Definition of electrical/electronic interface between the tether and the user bus.

CONTACTS:
- Marino Dobrowolny
- Les Johnson
- Joseph Kolecki
- Jim McCoy
- Juan Sanmartin
- Nobie Stone

REFERENCES:

Proc. of Fourth International Conference on Tethers in Space, Washington DC, 10-14 April 1995

-- ELECTRODYNAMICS --

ULF/ELF/VLF Communications Antenna

APPLICATION: Generation of ULF / ELF / VLF waves by an orbiting electrodynamic tether for worldwide communications.

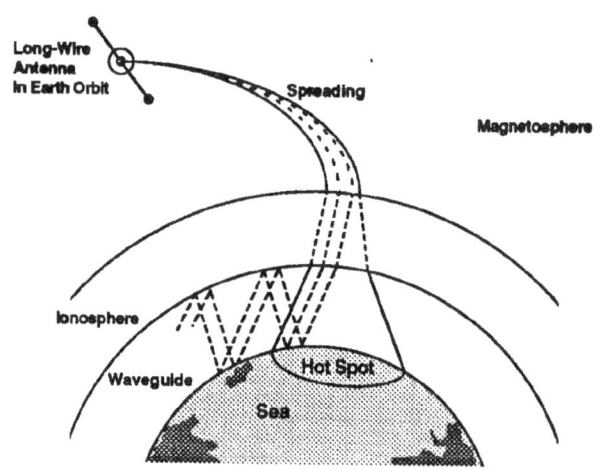

DESCRIPTION: An insulated conducting tether connected to a spacecraft, and terminated at both ends with plasma contactors. Variations in tether current can be produced to generate ULF/ELF/VLF waves for communications. This tether antenna can be self-powered (using the current induced in it by the geomagnetic field for primary power) or externally powered (fed by an on-board transmitter).

CHARACTERISTICS:
- Length: 20-100 km
- Tether Current: 10 A
- Potential For Technology Demonstration: Near-Term

CRITICAL ISSUES:
- Characterization of the transmitter
- Characterization of the propagation media (including the ionosphere at LEO altitudes, the lower atmosphere, and ocean water)
- Analysis of the sources of background noise and the statistical structure of that noise at the receiver
- Characterization of the instabilities and wave due to large current densities in the Alfven wings
- More advanced mathematical models are required for an adequate understanding of tether antenna systems, including the need to supersede the present cold-plasma based models with more accurate warm-plasma based models
- Determination of optimum ground station locations, including the possibility of mobile receivers
- Correlation of signals received at different ground station locations to subtract out noise

STATUS:
- TSS-1 and TSS- 1R flights

DISCUSSION: When a current flows through the tether, electromagnetic waves are emitted, whether the current is constant or time-modulated. The tether current can be that induced by tether motion through the geomagnetic field, or one generated by an on-board transmitter. Modulation of the induced current can be obtained by varying a series impedance, or by turning an electron gun on the lower end on and off, at the desired frequency. Waves are emitted by a loop antenna composed of the tether, magnetic field lines, and the ionosphere.

ULF/ELF/VLF waves produced in the ionosphere will be injected into the magnetosphere more efficiently than those from present ground-based man-made sources. These waves may provide instant worldwide communications by spreading over most of the Earth via the process of ducting. With a 20-100 km tether and a wire current of the order of 10 A, it appears possible to inject into the Earth-ionosphere transmission line power levels of the order of 1 W by night and 0.1 W by day.

CONTACTS:
- Robert Estes
- Mario Grossi
- Giorgio Tacconi

REFERENCES:

Grossi, M. D., "A ULF Dipole Antenna on a Spaceborne Platform of the PPEPL Class," Report for NASA contract NAS8-28203, May, 1973.

P.R.Bannister et al. "Orbiting Transmitter and Antenna for Spaceborne Communications at ELF/VLF to Submerged Submarines", Agard Conference Proceedings 529, May 1993, pp. 33-1-33-14

Proc. of Fourth International Conference on Tethers in Space, Washington DC, 10-14 April 1995

-- PLANETARY --

Aerocapture with Tethers for Planetary Exploration

APPLICATION: May provide significant mass savings when used in the exploration of the atmosphere-bearing planets and satellites in the solar system.

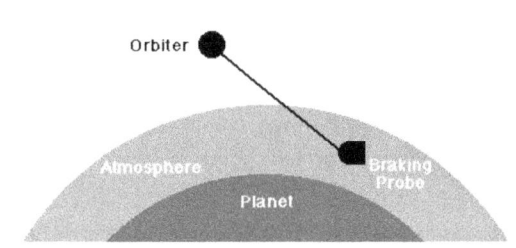

DESCRIPTION: The basic concept involves an orbiter and a probe connected by a long, thin tether. The probe is deployed into the atmosphere of a planet where aerodynamic drag decelerates it from hyperbolic approach speed to capture speed. The tension on the tether provides the braking effect on the orbiter, thus eliminating the need for a retro-propulsion maneuver. During the maneuver the orbiter travels outside the atmosphere and does not require heat shielding.

CHARACTERISTICS:

- Tether Length: 10-100 km
- Tether Diameter: 0.5-1.5 mm
- Orbiter Mass: 1000 kg
- Probe Mass: 1000 kg
- Probe Area: 500-3000 m^2
- Potential for Technology Demonstration: Mid-term

CRITICAL ISSUES:

- Reducing the probe area without causing significant bending in the tether.
- Assessing the effect of parameter uncertainties (such as atmospheric density, target altitude, ballistic coefficient and spin rate) on tether and maneuver design.
- Developing guidance and control laws and mechanisms to handle these uncertainties.

STATUS:

- Preliminary analyses demonstrate the feasibility of the concept.
- Reentry of SEDS-1 provides insight into the dynamics of a tether in an atmosphere.

DISCUSSION: Analytical and numerical studies have considered the possibility of using the aerobraking tether for the exploration of Venus, Mars, Jupiter, Saturn, Uranus, Neptune and Titan as well as for returning to Earth from Mars. One study compares the propellant mass of a typical rocket propulsion system to the tether mass required for the aerobraking system. In every instance in this study, the tether mass turns out to be less than the propellant mass.

The feasibility of the design is supported by studies that include flexibility, out-of-plane effects and parameter uncertainties. As a passive system, the aerobraking tether is less sensitive to parameter uncertainties than the typical aerobraking configuration.

For precise guidance, the system seems well suited to feedback control by adjusting the tether length.

CONTACTS:
- James M. Longuski
- Jordi Puig-Suari
- Steven G. Tragesser

REFERENCES:

Puig-Suari, J., "Aerobraking Tethers for the Exploration of the Solar System," Ph.D. Thesis, School of Aeronautics and Astronautics, Purdue University, West Lafayette, IN, August 1993.

Proc. of Fourth International Conference on Tethers in Space, Washington DC, 10-14 April 1995

-- PLANETARY --

Comet/Asteroid Sample Return

APPLICATION: Collection and
return to Earth of comet or asteroid
samples.

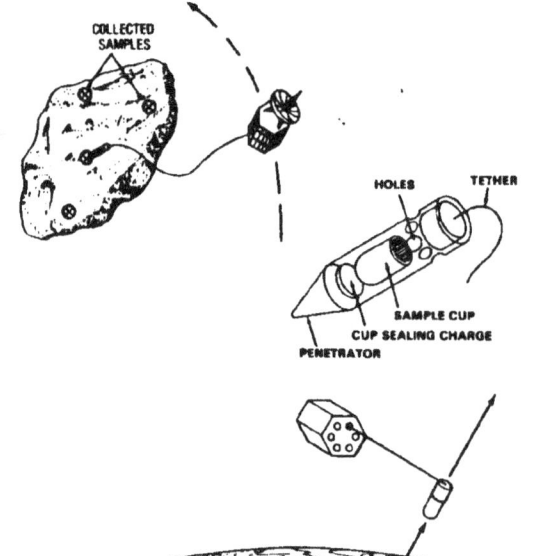

DESCRIPTION: Tethered
penetrators are launched from a
spacecraft during its rendezvous with a
comet or asteroid. They penetrate the
body's surface, collecting samples of
surface material. They are then reeled
aboard the spacecraft for return to
Earth. Using several penetrators,
samples could be collected from
different spots on one body, or from
more than one body.

CHARACTERISTICS:
- Tether Length: 50-100 m
- Tether System: Single Reel
- Penetrator System:
  Multiple Chambered
  Turret
- Penetrators:Core Drilling and
  Surface
- Deployment: Spring and Solid Rocket
- Potential for Technology
  Demonstration: Far-Term

CRITICAL ISSUES:
- Long-range, remote-controlled maneuvering and rendezvous
- Design and development of the penetrators, tether-reel subsystem, and
  penetrator turret subsystem

STATUS:
- Preliminary definition of the mission and hardware has been performedat
  JPL
- Detailed Analysis and design performed by Alenia for ESA's ROSETTA

DISCUSSION: The conventional approach to collecting samples from comets and
asteroids would be for a spacecraft to rendezvous with them and release a lander. The
lander would attach itself to the body in some way, drill for a core sample, and return to the
spacecraft. The sample would then be returned to Earth. A typical scenario would require
the following capabilities: (1) close range verification of a suitable landing and drilling site;
(2) automated and highly accurate soft landing; (3) lander attachment to the body (since
some would have very low gravity); (4) a drill unit with sufficient power to core a sample;

(5) lander separation from the body; (6) automated rendezvous with the orbiter; (7) sample transfer; (8) launch stage ejection; and (9) Earth return.

A tether approach would consist of the following sequence of events: (1) the spacecraft rendezvous with the comet or asteroid; (2) a tethered penetrator is shot at the target from a 50-100 m altitude; (3) on impact, sample material enters holes in the penetrator shell and fills the sample cup inside; (4) an explosive seals the cup and ejects it from the penetrator shell; (5) the cup velocity creates a tension in the tether as it rotates it; (6) spacecraft thrusters control the cup retrieval as it is reeled aboard; (7) other tethered penetrators retrieve samples from other areas or bodies; and (8) the spacecraft returns the samples to Earth.

In addition to the penetrator design described above, another type, in which the penetrator contains a core drill, could also be used. For this version, flanges would be extended upon impact, to secure the penetrator shell to the surface while the core sample is being drilled. The surfaces hardness would determine which type to use. Both types could be launched from the spacecraft by a spring and then propelled by attached solid rockets to the impact point. (This should impart sufficient momentum to permit a good surface penetration.) To allow a single tether reel subsystem to handle many penetrators, a rotatable turret with multiple, chambered penetrators could be used.

This tether system has the advantage of being simpler than a lander system (not requiring many of the capabilities listed for a lander system), and of allowing the collection of samples from more than one spot or body. The cost of such a tether mission has been estimated to be about $750 M, as opposed to about $1-2 B for a lander mission. However, the two methods are complementary in that the lander provides a single very deep sample and the penetrator provides smaller samples from different areas or bodies.

CONTACTS:
- Pietro Merlina
- Paul Penzo

REFERENCES:

"Tether Assisted Penetrators for Comet/Asteroid Sample Return," by Paul A. Penzo (JPL); paper presented at 1986 AIAA/AAS Astrodynamics Conference.

"Feasibility Assessment of a Tethered Harpoon for the ROSETTA backup Sampling", Alenia Spazio, SD-RP-AI-040, January 1990

"CSNR, Mission and System Definition Document", ESA SP-1125, June 1991

Jupiter Inner Magnetosphere Maneuvering Vehicle

APPLICATION: Generation of electromagnetic thrust or drag for maneuvering within
the inner Jovian magnetosphere.

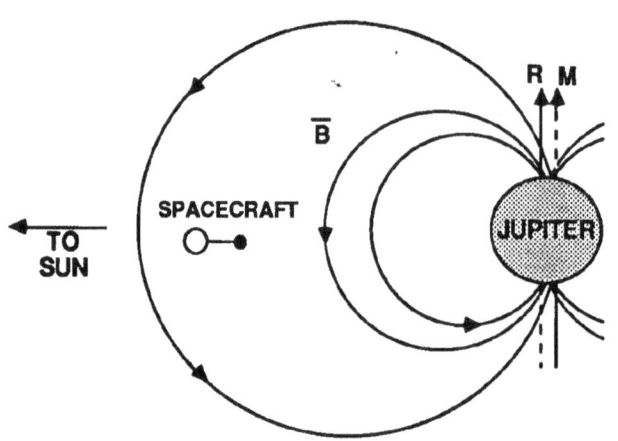

DESCRIPTION: An insulated conducting
tether connected to a spacecraft and possibly
terminated with a subsatellite. Plasma contactors
are used at both tether ends. When used
selectively with an on-board power supply
(probably nuclear) or a load, it interacts with the
Jovian magnetic field to produce thrust, drag and
electrical power as required to change orbital
altitude or inclination.

CHARACTERISTICS:
  •   Physical Characteristics: Undetermined
  •   Potential For Technology Demonstration: Far-Term

CRITICAL ISSUES:
  •   Successful operation of hollow cathodes or related active collectors
           as plasma contactors
  •   Assurance of long-term insulator life
  •   Susceptibility to micrometeoroid/debris damage
  •   Successful operation of a power supply (probably nuclear) with
           sufficient output power density
  •   Characterization of the performance of an electromagnetic tether
           in the Jovian Magnetosphere

STATUS:
  •   TSS-1, demonstrating electrodynamic applications, is scheduled for
           a 1991 launch
  •   No detailed system design study for this application has been
           performed

DISCUSSION: Since Jupiter's magnetic field is about twenty times that of Earth, an
electromagnetic tether should work well there. Because of Jupiter's rapid rotation
(period = 10 hrs), at distances greater than 2.2 Jovian radii from its center, the
Jovian magnetic field rotates faster than would a satellite in a circular Jovian orbit.
At these distances, the magnetic field would induce an emf across a conducting
tether, and the dissipation of power from the tether would produce a thrust (not
drag) on the spacecraft/tether system. At lesser distances, the satellite would rotate
faster than the magnetic field, and dissipation of tether power would produce drag
(not thrust). Examples of induced tether voltages are:
-10 kV/km (for drag) in LJO; and +108, 50, 21, and 7 v/km (for thrust) at Io,
Europa, Ganymede, and Callisto, respectively.

Inside the Jovian magnetosphere, at distance > 2.2 Jovian radii, the spacecraft could decrease altitude (decelerate) by feeding power from an on-board power supply into the tether against the induced emf. Below 2.2 radii, power from the tether could be dissipated. To return to higher altitudes, the process could be reversed.

Since the gravitational attraction of Jupiter is so strong, the energy required to descend to (or climb from) a very low Jupiter orbit is prohibitive for any conventional propulsion system. To descend to the surface of Jupiter from a distance of, say, 100 Jovian radii, an energy density of a little over 200 kW-hr/kg would be required for propulsion. Using this as a conservative estimate of the required performance of a tether system, it should be well within the capability of a nuclear power supply.

Recommendations were made at the Tether Workshop in Venice (October 1985) for a Jupiter inner magnetosphere survey platform to operate in the range from one to six Jovian radii. The electromagnetic tether in this application would be used primarily for orbital maneuvering. It could also assist a Galileo-type satellite tour (all equatorial), sampling of the Jovian atmosphere, and rendezvous with a Galilean satellite.

CONTACTS:
- Paul Penzo
- James McCoy

REFERENCES:
Applications of Tethers in Space, NASA CP-2422, March 1986.

Gabriel, S. B., Jones, R. M., and Garrett, H. B., "Alfven Propulsion at Jupiter," Tether Int. Conf. 1987.

Penzo, P. A., "A Survey of Tether Applications to Planetary Exploration," AAS 86-206, AAS Int. Conf. 1986.

-- PLANETARY --

Mars Tethered Observer

APPLICATION: Provide instrument access to low orbital altitudes for periodic *in-situ* analysis of the upper Martian atmosphere.

DESCRIPTION: An instrument package attached by a deployable tether (up to 300 km in length) to an orbiting Mars Observer spacecraft.

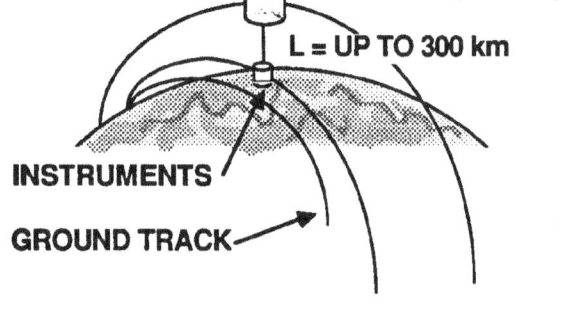

CHARACTERISTICS:
- Length: Up to 300 km (Tether is not vertical)
- Satellite Altitude: 350 km
- Instrument Altitude: Down to 90 km
- Potential For Technology Demonstration: Mid-Term

CRITICAL ISSUES:
- Tether material (graphite is a potential candidate) and Orbiter fuel consumption

STATUS:
- System performance analysis for various altitudes and different mission scenarios of the probe performed by the Smithsonian Astrophysical Observatory

DISCUSSION: The purpose of the mission itself is to analyze the composition and chemistry of the Martian atmosphere for one Martian year. The tether would allow instruments to be lowered periodically for *in-situ* measurements at lower altitudes and collection martian dust during storms thus saving on landers's costs. A tether (Up to 300 km long) could be used with the observer as it orbits Mars at an altitude of 350 km. The instrument package would be deployed for a few hours at a time, perhaps every two months, or so. Additional propulsion capability would be required for the observer for altitude maintenance. Although addition of the tether system would increase the mission cost, it should greatly enhance its scientific value.

CONTACTS:
- Enrico Lorenzini
- Paul Penzo
- Monica Pasca

REFERENCES:

Proc. of Fourth International Conference on Tethers in Space, Washington DC, 10-14 April 1995

Lorenzini, E.C., MD, Grossi, and M. Cosmo, " Low Altitude Tethered Mars Probe," Acta Astronautica, Vol 21, No.1, 1990, pp. 1-12.

Pasca M. and E. C. Lorenzini, "Optimization of a Low Alitude Tethered Probe for Martian Atmospheric Collection", The Journal of the Astronautical Sciences, Vol. 44, No.2, 1996, pp.191-205

-- PLANETARY --

Tethered Lunar Satellite for Remote Sensing

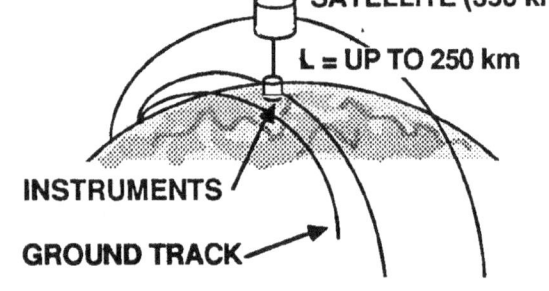

APPLICATION: Provide instrument access to low, unstable, lunar orbital altitudes.

DESCRIPTION: An instrument package at low altitude, suspended by a tether from a satellite in a higher, stable, polar orbit around the moon.

CHARACTERISTICS:
- Tether Length: 90 -250km

- Instrument Altitude: up to 50 km

- Potential For Technology Demonstration: Far-Term

CRITICAL ISSUES:
- Assurance of acceptable strength and flexibility for the tether material
- Susceptibility to micrometeoroid/debris damage

STATUS:
- PROTEUS (PRObe Tethered for Exploration of Uncovered Satellites) study performed by ALENIA Spazio. Analysis of mission scenarios and scientific objectives

DISCUSSION: Due to Sun and Earth perturbations, close lunar satellites would be unstable and short lived (perhaps a few months). However, as proposed by Giuseppe Colombo, access to low lunar orbits could be achieved by tethering an instrument package to a satellite in a stable lunar orbit. The package could be lowered as close to the Moon as desired. One proposed configuration would tether an instrument package 50 km above the lunar surface from a satellite in a stable 300 km orbit. By using a polar orbit, complete coverage of the lunar surface could be obtained. Occasional adjustments to the tether length may be required to keep the package at a safe altitude. Sensitive measurements of lunar magnetic field and gravitational anomalies could be performed.

CONTACTS:
- Pietro Merlina
- Paul Penzo

REFERENCES:
Colombo G., et al., "Dumbbell Gravity Gradient Sensor: A New Application of Orbiting Long Tethers, SAO Report in Geoastronomy No. 2, June 1976

Merlina P, "PROTEUS-PRObe Tethered for Exploration of Uncovered Satellites: The Proteus Lunar Mission, ESA WPP-081, 1994, pp.512-527

-- SCIENCE --

Science Applications Tethered Platform

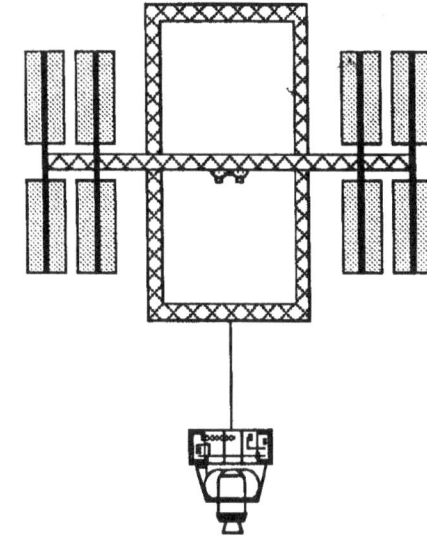

APPLICATION: Provides a remote platform to the Space Station for space and Earth observation purposes.

DESCRIPTION: A platform, attached to the Space Station by a multifunction tether (power link, data link), provides a new means to allow high precision pointing performance by the combination of disturbance attenuation via tether and active control of a movable attachment point.

CHARACTERISTICS:
* Length: 10 km
* Mass: 10,000 kg
* Power required: Up to
15 kW by Tether

Power Line Link
* Link Data Rate: Up to 20 Mb/s by Tether
Optical Fibers Link
* Pointing Accuracy: Up to 10 Arcseconds

* Potential For Technology Demonstration: Mid-Term

CRITICAL ISSUES:
* Space Station impacts
* Dynamic noise induced on tether
* Movable attachment point control
* Power link technology
* Optical fibers link technology
* Tether impact protection technology

STATUS:
* ASI/Aeritalia SATP Definition Study in initial design assessment phase, mid-term report issued in March 1986. Final report for the current study phase issued in May 1987
* Ball Aerospace, Selected Tether Applications Study Phase III

DISCUSSION: A tethered pointing platform would take advantage of the facilities of the station for maintenance and repair while being isolated from contamination and mechanical disturbances. As an initial step, a medium size pointing platform seems the most suitable facility for a class of observational applications. In fact, if ambitious astrophysical projects justify the design of a dedicated complex free-flyer, medium observational applications of relatively short duration could take advantage of a standard pointing facility able to arrange at different times several observational instruments. This pointing facility could allow reduction of costs, avoiding the cost of separate service functions for each application.

91

CONTACTS:
- Franco Bevilacqua
- Pietro Merlina
- James K. Harrison

REFERENCES:

Applications of Tethers in Space, NASA CP-2422, March 1986.

SATP Definition Study, Mid-Term Report, Aeritalia, TA-RP-AI-002, March 21, 1986.

SATP Definition and Preliminary Design, Final Report, Aeritalia, TA-RP-AI-006, 1987.

Proc. of Fourth International Conference on Tethers in Space, Washington DC, 10-14 April 1995

-- SCIENCE --

Shuttle Science Applications Platform

APPLICATION: Provides a remote
platform to the Space Shuttle for various science
and applications purposes.

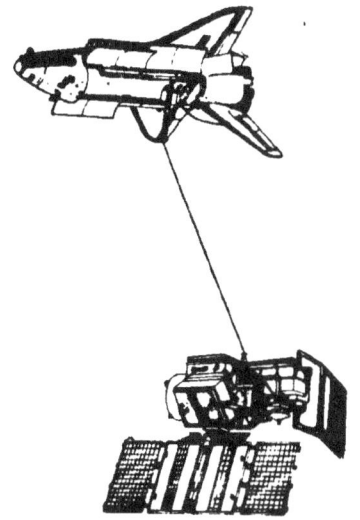

DESCRIPTION: A platform, attached to
the Space Shuttle by a tether, provides a unique
means by which remote applications may be
performed.

CHARACTERISTICS:
- •   Physical Characteristics: TBD
- •   Potential For
       Technology
       Demonstration: Near-Term

CRITICAL ISSUES:
- •   Dynamic noise induced on tether
- •   Micrometeoroid damage

STATUS:
- •   Various investigators (listed below) have examined preliminary concepts

DISCUSSION: Possible uses for a remote platform include stereoscopic sensing,
magnetometry, atmosphere science experiments, and chemical release experiments.

CONTACTS:
- •   Franco Angrilli
- •   Franco Bevilacqua
- •   Franco Mariani
- •   Antonio Moccia
- •   Sergio Vetrella

REFERENCES:
Applications of Tethers in Space, NASA CP-2422, March 1986.

Proc. of Fourth International Conference on Tethers in Space, Washington DC,
10-14 April 1995

-- SCIENCE --

Tethered Satellite for Cosmic Dust Collection

APPLICATION: To collect micrometeoric material from the upper atmosphere.

DESCRIPTION: A satellite tethered to the Space Shuttle is lowered into the upper atmosphere. The surface of the satellite contains numerous small collecting elements which would document the impact of cosmic dust or actually retain the particles for analysis back on Earth.

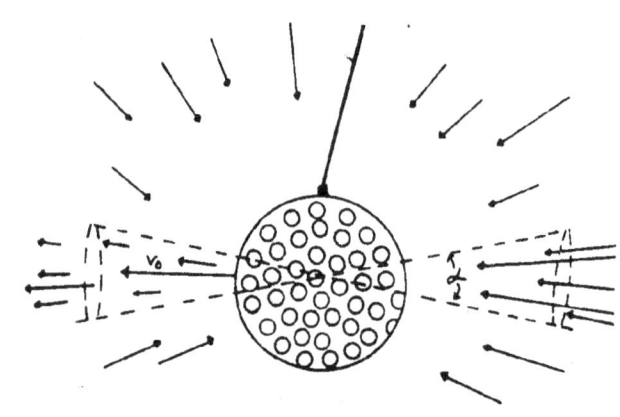

CHARACTERISTICS:
- Tether Length: 100 km
- Operating Altitude: 120 km
- Tether Diameter: 1 meter
- Power Requirements: Minimal, enough to operate solenoid activated irises
- Potential For Technology Demonstration: Near-Term

CRITICAL ISSUES:
- Efficient analysis of large collector surface areas to detect micron-sized particles and impact craters

STATUS:
- Preliminary concept design investigated at Indiana University Northwest

DISCUSSION: This concept proposes to collect intact cosmic dust particles smaller than 2 microns which impact the collector surface at velocities less than 3 km/sec, and the study of impact craters and impact debris which result from impacts of all sized particles at velocities greater than 3 km/sec. It is estimated that at a 120 km altitude, between 1×10^3 and 1×10^4 particles will survice collection intact per square meter per day, and between 2×10^4 and 2×10^5 impact craters will be recorded per square meter per day. The figure in the illustration above represents the "survivable" impact cones for particles striking a tethered satellite. For a maximum impact velocity of 3 km/sec, a is approximately 22 degrees.

CONTACTS:
- George J. Corso

REFERENCES:
G.J. Corso, "A Proposal to Use an Upper Atmosphere Satellite Tethered to the Space Shuttle for the Collection of Micro-meteoric Material," Journal of the British Interplanetary Society, Vol. 36, pp. 403-408, 1983.

-- SPACE STATION --

Microgravity Laboratory

APPLICATION: Provide a readily accessible
laboratory in Earth orbit with the minimum gravity
level possible.

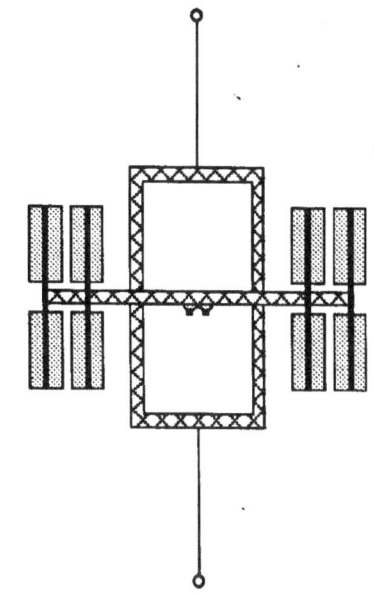

DESCRIPTION: A laboratory
facility on board the Space Station at its
vertical center of gravity. Two opposing
tethers with end masses are deployed
vertically from the Space Station (one
above and one below). Their lengths are
varied to control the Space Station center
of gravity, placing it on the microgravity
modules to minimize their gravity
gradient acceleration (artificial gravity
level).

CHARACTERISTICS:
- Physical Characteristics: TBD

CRITICAL ISSUES:
- Evaluation of the overall impacts to the Space Station
- Determination of just how good the lab's microgravity would be
- Identification of the process and technologies to be studied in microgravity, and the laboratory facilities and capabilities they will require
- Development of the necessary gravity-measuring instrumentation
- Evaluation of the tether system's cost effectiveness

STATUS:
- A JSC tethered gravity laboratory study (addressing the issues of active center-of-gravity control, identification of low-gravity processes to be studied, and evaluation of the laboratory g-level quality)
- SEDS-1 and -2 missions abd TSS-1 and TSS-1R have provided measurements of the acceleration fields and associated noise during tether and payload deployment

DISCUSSION: To allow the performance of experiments under microgravity conditions
(10^{-4} g and less) for extended periods of time, a microgravity laboratory facility could be
incorporated into the Space Station. The laboratory modules would be located on the Space
Station proper, at its center of gravity. Two opposing TSS-type tethers with end masses
would be deployed vertically from the Space Station (one above and one below), to assure
that the station center of gravity is maintained within the lab modules. Its exact location
would be controlled by varying the upper and lower tether lengths, allowing prolonged and
careful control of the residual microgravity magnitude and direction inside the lab. A
nearly constant microgravity could be maintained. These tethers would lower the gravity-
gradient disturbances transmitted to the experiments being performed while enhancing

95

station attitude control. Although people would be a major source of disturbances, human access to microgravity experiments is preferred (at least initially) over remote access. This configuration would easily accommodate this preference.

One candidate microgravity lab currently under study for the Space Station, is the Materials Technology Lab (MTL). It is projected to be a common module, equipped as a lab, to perform a variety of experiments related to materials technology. Biological experiments may also be performed in microgravity in another module.

Although this is the preferred microgravity lab configuration, two alternatives are also possible. One would be to have the lab connected by a crawler to a single tether from the Space Station. The crawler would position the lab on the station-tether system center of gravity. The other configuration would be to fix the lab to a single tether from the station. The lab would be positioned at the system center of gravity by varying the tether length. Both alternatives have the advantage of isolating the lab from disturbances, but they have the disadvantages of reducing human access and probably precluding the use of the microgravity modules planned for the initial Space Station.

CONTACTS:
- Franco Bevilacqua
- Mario Cosmo
- Pietro Merlina
- Enrico Lorenzini

REFERENCES:

Applications of Tethers in Space, NASA CP-2422, March 1986. (pp. 223-238)

G. Von Tiesenhausen, ed., "The Roles of Tethers on Space Station," NASA TM-86519, Marshall Space Flight Center, October 1985.

Lorenzini, E.C., "A Three-Mass Tethered System for Micro-g/Variable-g Applications," Journal of Guidance, Control, and Dynamics, Vol. 10, No.3, May-June 1987, pp. 242-249

-- SPACE STATION --

Shuttle Deorbit from Space Station

APPLICATION: Allows the Shuttle Orbiter to be deboosted to Earth while the Space Station is boosted to a higher orbit.

DESCRIPTION: Upon completion of a Shuttle re-supply operation to the Space Station, the Shuttle is deployed on a tether toward the Earth. The Space Station, accordingly, is raised into a higher orbit, causing excess momentum to be transferred from the Shuttle orbit to the Space Station orbit. After deployment, the Shuttle is released causing the Shuttle to deorbit.

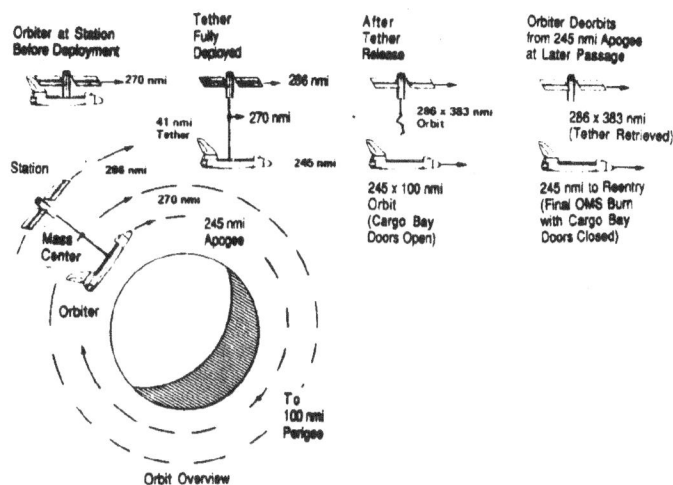

CHARACTERISTICS:
- Initial Space Station/Shuttle Orbit: 500 km
- Tether Length: 65 km
- Final Space Station Orbit: 518 x 629 km
- Final Shuttle Orbit: 185 x 453 km
- Estimated Mass: 250,000 kg (Space Station) 100,000 kg (Shuttle)
- Potential For Technology Demonstration: Mid-Term

CRITICAL ISSUES:
- Excess angular momentum scavenged by Space Station must be used in order to beneficially use this application
- Dynamic noise induced by tether deployment and separation
- Alignment of tether to Space Station to eliminate torques

STATUS:
- Martin Marietta, Selected Tether Applications Study, Phase III
- NASA-MSFC System study

DISCUSSION: This application potentially could be one of the most cost effective uses of a tether. The main disadvantage is that the excess momentum transferred to the Space Station must be efficiently used, otherwise the station will be in an orbit too high for subsequent Shuttle re-supply missions. Several ideas on use of this excess momentum have been studied, such as altering STV boosts by the Space Station with Shuttle re-supply missions (see Application "Tethered STV Launch"). Another method is using an electrodynamic tether (see Application "Electrodynamic Power Generator") to generate power at the expense of orbital energy to deboost the Space Station.

CONTACTS:
- James K. Harrison
- Les Johnson

97

REFERENCES:

G. Von Tiesenhausen, ed., "The Roles of Tethers on Space Station," NASA TM-86519, Marshall Space Flight Center, October 1985.

-- SPACE STATION --

Tethered STV Launch

APPLICATION: Allows an STV to be boosted to a higher orbit at the expense of Space Station angular momentum.

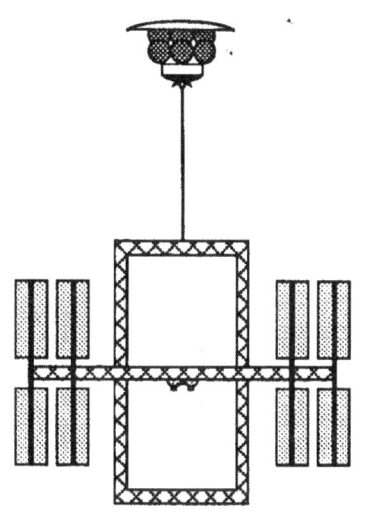

DESCRIPTION: An STV would be deployed from the Space Station on a tether away from Earth, in preparation for launch. Upon separation from the tether, orbital angular momentum is transferred from the Space Station to the STV, causing the Space Station Altitude to be lowered while that of the STV is raised.

CHARACTERISTICS:
- Initial Space Station/
  STV Orbit: 500 km
- Tether Length: 150 km
- Final Space Station
  Orbit: 377 x 483 km
- Final STV Orbit: 633 x 1482 km
- Estimated Masses: 250,000 kg
  (Space Station)
  35,000 kg (STV)

- Potential For Technology Demonstration: Far-Term

CRITICAL ISSUES:
- Angular momentum taken away from the Space Station must be resupplied in order to beneficially use this application
- Dynamic noise induced by tether deployment and separation
- Alignment of tether to Space Station to eliminate torques

STATUS:
- Martin Marietta, Selected Tether Applications Study Phase III

DISCUSSION: Martin Marietta has studied the application of tethered deployment of the STV as well as Shuttle from the Space Station. Either of these applications alone would cause an unacceptable change in altitude of the Space Station. When combined, properly sequencing STV launches and Shuttle deorbits, the orbital angular momentum of the Space Station may be preserved while providing a large net propellant savings for the Shuttle, STV and Space Station.

99

CONTACTS:
- James K. Harrison
- Les Johnson

REFERENCES:

Applications of Tethers in Space, NASA CP-2422, March 1986.

G. Von Tiesenhausen, ed., "The Roles of Tethers on Space Station," NASA TM-86519, Marshall Space Flight Center, October 1985.

Application "Shuttle Deorbit From Space Station"

Proc. of Fourth International Conference on Tethers in Space, Washington DC, 10-14 April 1995

-- SPACE STATION --

Variable/Low Gravity Laboratory

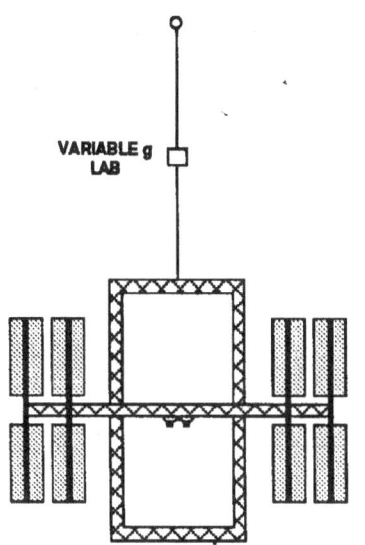

VARIABLE g
LAB

APPLICATION: Provide a readily accessible laboratory in Earth orbit with a variable, low-gravity level.

DESCRIPTION: A laboratory facility, attached by a crawler to a tether deployed vertically from the Space Station. The gravity gradient between the station-tether system center of gravity and the laboratory produces an artificial-gravity force throughout the lab. The lab gravity level, with a constant vertical direction, is varied by changing the lab and crawler distance from the system's center of gravity. The lab can attain microgravity levels if it can move to the center of gravity.

CHARACTERISTICS:
- Physical Characteristics: TBD
- g-Level: Up to 10^{-1}
- Potential For Technology Demonstration: Far-Term

CRITICAL ISSUES:
- Evaluation of the overall impacts to the Space Station
- Determination of just how good the lab's low gravity would be
- Identification of the processes and technologies to be studied in low gravity, and the laboratory facilities and capabilities they will require
- Development of the necessary gravity-measuring instrumentation
- Evaluation of the tether system's cost effectiveness
- Determination of how gravity-level medical experiments should be performed in a Space Station system
- Design of a tether crawler and lab module
- Development of systems for the remote control of the lab experiments

STATUS:
- A study by Alenia-SAO-Padua U. for NASA-JSC on tethered gravity laboratory study (addressing the issues of active center-of-gravity control, identification of low-gravity processes to be studied, and evaluation of the laboratory g-level quality)
- A study by SAO for NASA-MSFC on tethered variable gravity elevators.
- TSS-1 and -1R, SEDS-1 and -2 have provided measurements of the acceleration field change and associated noise during tether and payload deployment

101

DISCUSSION: To allow the performance of experiments under conditions of constant or variable low gravity (up to 10^{-1} g) for extended periods of time, a variable/low gravity lab could be attached to a crawler on a tether deployed vertically from the Space Station. The artificial gravity at any point along the tether is produced by the gravity gradient between that point and the station/tether system center of gravity, and is proportional to the distance between them. The lab could vary its gravity level, with a constant direction, by varying its distance from the system center of gravity. A constant gravity level could be maintained by adjusting the lab position to compensate for orbital variations in the system gravity level. The lab could also attain microgravity levels if it could move to the center of gravity. This lab could study processes with both gravity and time as variables. It has been calculated the the lab could attain g-levels of 10^{-6}, 10^{-4}, 10^{-2}, and 10^{-1} at distances above the center of gravity of about 2 m, 200 m, 20 km, and 200 km, respectively.

In addition to easy gravity control, the use of a tether system for a low gravity lab would have other advantages. It would reduce disturbances transmitted to the lab (to about 10^{-8} g), minimize the gravity gradient acceleration inside the lab, and enhance overall system attitude control. It would have the disadvantage of reducing human access to lab experiments, requiring the increased use of remote controls. Also, it could only provide a gravity level of up to 10^{-1} g.

This lab could be used to examine the effects of low gravity on both physical and biological processes. Some biological processes of interest would be plant and animal growth, and human performance and medical processes (such as those related to the cardiovascular, skeletal, and vestibular systems). Such physical processes as crystal growth, fluid science, and chemical reactions could be studied. Conditions on low gravity bodies (such as asteroids) could be simulated to examine natural processes (such as meteor impacts). Of particular interest would be the determination of the gravity threshold for various processes.

CONTACTS:
- Chris Rupp
- Silvio Bergamaschi
- Franco Bevilacqua
- Mario Cosmo
- Enrico Lorenzini
- Pietro Merlina

REFERENCES:
Applications of Tethers in Space, NASA CP-2422, March 1986.

G. Von Tiesenhausen, ed., "The Roles of Tethers on Space Station," NASA TM-86519, Marshall Space Flight Center, October 1985.

F. Bevilacqua and P. Merlina, "The Tethered Space Elevator System," Second International Conference on Tethers In Space, Venice, Italy, 1987.

E.C. Lorenzini et al., "Dynamics and Control of the Tether Elevator Crawler System", Journal of Guidance, Control, and Dynamics, Vol. 12, No.3, pp. 404-411 1989

-- SPACE STATION --

Attitude Stabilization and Control

APPLICATION: Provides the Space Station with restoring torques around pitch and roll axes

DESCRIPTION: A tethered ballast could be deployed to serve as an attitude stabilizer. This feature could be used on a temporary basis during the construction of the Space Station or on a permanent basis to alleviate the CMG's requirements as well as function as a backup facility in case of ACS failure.

CHARACTERISTICS:
- Mission Duration: up to some days
- Masses: Deployer ~ 650 Kg;
  Tether ~ 400 Kg;
  Ballast ~ 1400 Kg
- Tether Length: 6000 m
- Potential For
  Technology
  Demonstration: Mid-Term

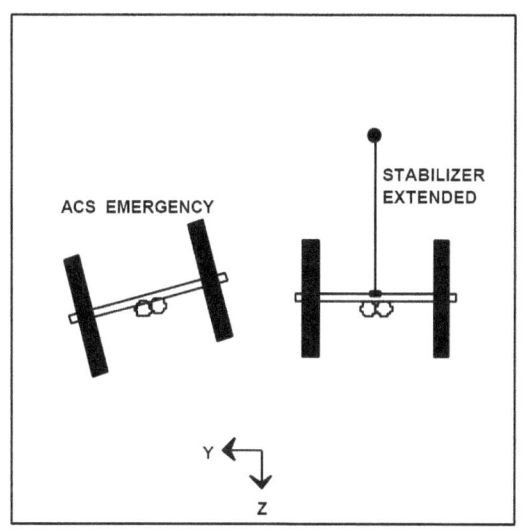

CRITICAL ISSUES:
- Attitude dynamics of the tether-stabilized Station during -deployment of ballast.
- Assessment of mass, propellant, CMG's sizing, redundancy philosophy and contingencyreboost scenario.

STATUS:
- Feasibility analysis performed by Alenia and SAO for NASA/JSC

DISCUSSION: The typical configuration of the Space Station results in a spacecraft that requires a complex and careful design of the Attitude Control System. CMG's sizing and RCS propellant allocated depend on several nominal and emergency operations that need to be managed. The attitude tether stabilizer concept seems to have the potential for being an effective way of overcoming some of the above difficulties The advantages include: system simplicity, relatively low costs and reusability.

CONTACTS:
- Pietro Merlina
- Enrico Lorenzini

REFERENCES:
"Tethered Gravity Laboratories Study". Performed by ALENIA Spazio and SAO under NASA-JSC Contract NAS9-17877.

Generalized Momentum Scavenging from Spent Stages

APPLICATION: Scavenge angular momentum from a spent stage for the benefit of the payload.

DESCRIPTION: After the injection of an upper stage and its payload into an elliptical park orbit, the payload is tethered above the spent stage. At the proper time, the payload is released which causes a payload boost and spent stage deboost.

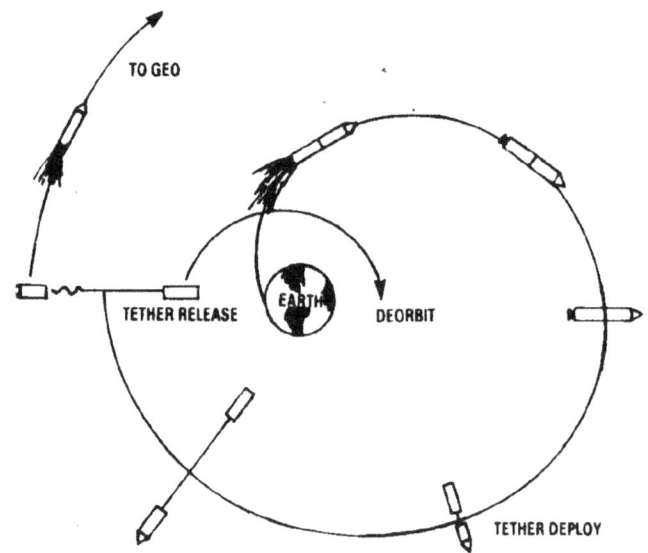

CHARACTERISTICS:
- Physical Characteristics : TBD
- Potential For
  Technology
  Demonstration: Mid-Term

CRITICAL ISSUES:
- Mass of tether and reel equipment versus payload performance gain
- Integration impact on systems

STATUS:
- Preliminary evaluation completed by MIT , Michoud and Tether Applications
- Detailed analysis in progress at SAO in collaboration with Tether Unlimited

DISCUSSION: This concept appears to be impractical due to mass relationships and integration costs. The most immediate application is for newly developed upper stage/payload combinations and those having a high ratio of spent upper stage to payload mass.

CONTACTS:
- Manual Martinez-Sanchez
- Joe Carroll
- Les Johnson
- Enrico Lorenzini

REFERENCES:

J.A. Carroll "Guidebook for Analysis of Tether Applications," Contract RH4-394049, Martin Marietta Corporation, March 1985. Available from the author

M. Martinez-Sanchez, "The Use of Large Tethers for Payload Orbital Transfer," Massachusetts Institute of Technology, 1983.

G. Colombo, "The Use of Tethers for Payload Orbital Transfer," NASA Contract NAS8-33691, SAO, Vol. II, March 1982.

-- TRANSPORTATION --

Internal Forces for Orbital Modification (Orbital Pumping)

APPLICATION: To change the orbital eccentricity of a Space Station or platform without the use of propulsion systems.

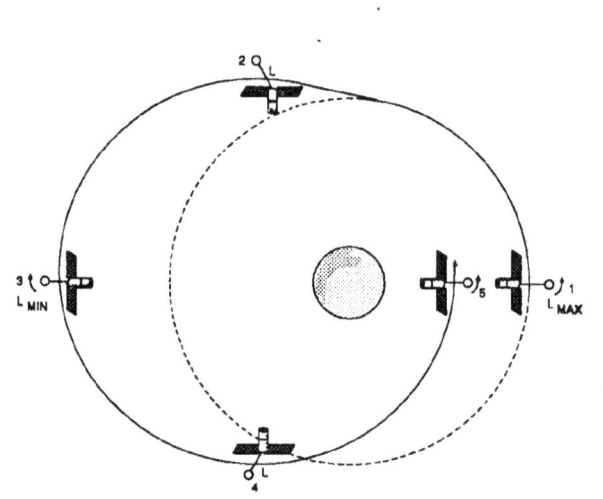

DESCRIPTION: The internal mechanical energy of a Space Station (in the form of excess electrical energy transferred to a motor) is used to vary the length of a tether attached to an end mass. The length is changed in phase with the natural libration of the tether, which is known as libration pumping. Proper timing of tether deployment and retrieval done in this fashion can be used to change the orbital eccentricity.

CHARACTERISTICS:
- Physical Characteristics: Undetermined
- Potential For
  Technology
  Demonstration: Mid-Term

CRITICAL ISSUES:
- Internal vs. external energy trade-off
- Power required and heat generated by the operation
- Change in orbits is relatively slow

STATUS:
- Preliminary feasibility shown by Martin Marietta Denver

DISCUSSION: Orbit eccentricity can be increased by libration pumping as is shown in the illustration. At (1) the mass is fully extended, and libration starts. At (2), with the mass in a prograde swing, the retrieval motor pulls the spacecraft toward the mass, adding energy to the orbit. At (3), which is the new apogee of the orbit, the tether length is at a minimum. At (4), with the mass in a retrograde swing, the tether is re-deployed and the retrieval brakes are used to dissipate orbital energy in the form of excess heat. At (5), the new perigee, the mass is again fully deployed. This procedure is repeated until the desired eccentricity is reached.

CONTACTS:
- Manual Martinez-Sanchez
- Joe Carroll

REFERENCES:

G. Von Tiesenhausen, ed., "The Roles of Tethers on Space Station," NASA TM-86519, Marshall Space Flight Center, October 1985.

Breakwell, J. V., Gearhart, J. W., "Pumping a Tethered Configuration to Boost its Orbit Around an Oblate Planet," AAS 86-217, Int. Conf. 1986.

Satellite Boost from Orbiter

APPLICATION: Boost a satellite payload into a circular or elliptical orbit higher than the Orbiter orbit.

DESCRIPTION: A satellite is deployed along a tether "upward" (away from the Earth) from the Shuttle Orbiter. Libration begins and momentum is transferred from the Shuttle orbit to the satellite. The satellite is released and placed into a higher orbit while at the same time giving the Shuttle a deboost to return to Earth. Less fuel is required for both the satellite and the Orbiter. A TSS-derived deployer could be used.

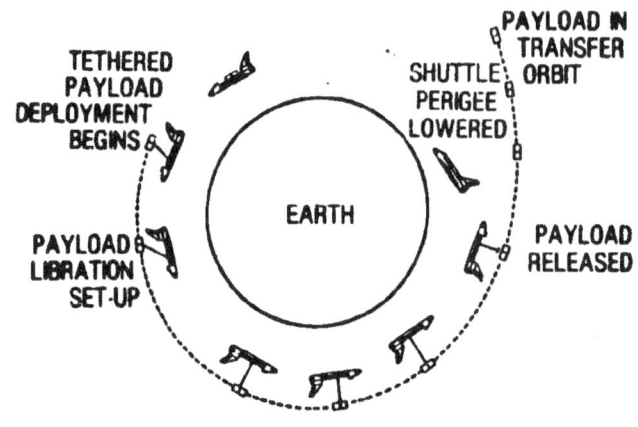

CHARACTERISTICS:
- Length: Dependent on desired orbit (see "Discussion" below)
- Tether System: Either permanent or removable from Orbiter
- Potential For Technology Demonstration: Near-Term

CRITICAL ISSUES:
- Release mechanism for payload
- Airborne support equipment for Orbiter
- Micrometeorite damage

STATUS:
- Energy Science Lab development contract completed March 1987
- MIT, Martin Marietta-Denver have completed preliminary assessment
- Ball Aerospace, Selected Tether Applications Study, Phase III
- SAO analysis for "SEDSAT" mission

DISCUSSION: This application has been studied in various forms by several contractors as noted above. One example studied is the tethered deployment of the AXAF (Advanced X-Ray Astrophysics Facility) into its operational orbit. For this example, the AXAF is assumed to have a mass of 9,070 kg and the Shuttle (after deployment) a mass of 93,000 kg. With the Shuttle and AXAF at an initial elliptical orbit of 537 x 219 km, the AXAF is deployed along a 61 km tether. As momentum is transferred from Shuttle to AXAF, the Shuttle orbit descends to a new 531 x 213 km and the AXAF orbit ascends to a new 593 x 274 km orbit. After tether separation, the AXAF is directly inserted into a 593 km circular orbit. Simultaneously, the Shuttle takes on an elliptical 531 x 185 km orbit, from which it will make a final OMS burn before its reentry.

CONTACTS:
- James K. Harrison
- Joe Carroll
- Les Johnson
- Enrico Lorenzini
- Manual Martinez-Sanchez

REFERENCES:

Applications "Upper Stage Boost from Orbiter" and "Small Expendable Deployer System"

Carroll, J. A., "Guidebook for Analysis of Tether Applications," Contract RH4-394049, Martin Marietta Corporation, Feb. 1985.

Proc. of Fourth International Conference on Tethers in Space, Washington DC, 10-14 April 1995

-- TRANSPORTATION --

Shuttle Docking by Tether

APPLICATION: Enables Shuttle Orbiter to dock to other structures such as the Space Station.

DESCRIPTION: A tether deployed by the Space Station is attached to a docking module. This module would capture and retrieve the Shuttle, allowing a remote rendezvous.

CHARACTERISTICS:
*   • Tether Length: 40-100 Km
*   • Potential For
        Technology
        Demonstration: Mid-Term

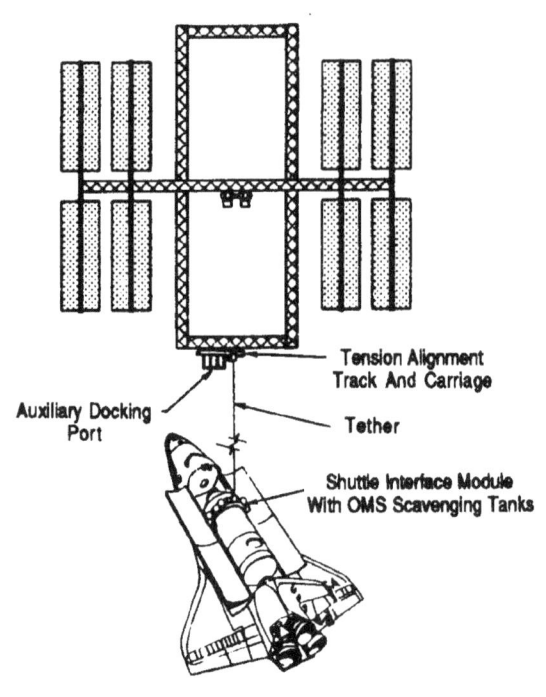

CRITICAL ISSUES:
* • Accurate guidance system, such as GPS needed
* • Rendezvous and capture technique definition required
* • Post-rendezvous tether dynamics
* • Alignment of tethertension with Station center of mass

STATUS:
* •   Martin Marietta, Selected Tether Applications Study, Phase III

DISCUSSION: A tether, attached to a docking module, would be deployed toward the Earth from the Space Station. The length of deployment is adjusted so that the velocity of the docking module matches the velocity at apogee of an elliptical orbit of the Shuttle. This would cause increased OMS propellant available to the Shuttle. This application would probably be combined with Application "Shuttle Deorbit from Space Station".

CONTACTS:
* •   James K. Harrison
* •   Chris Rupp

REFERENCES:
Applications of Tethers in Space, NASA CP-2422, March 1986.

-- TRANSPORTATION --

Tether Reboosting of Decaying Satellites

APPLICATION: To retrieve, repair, and reboost a
defective or decaying satellite.

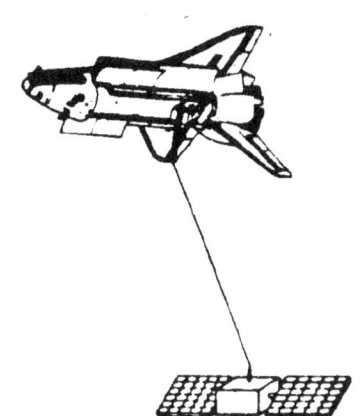

DESCRIPTION: A permanent tether attached to
the Space Shuttle is used to rendezvous with a decaying
satellite. It can then either be repaired by Shuttle
crewmen and/or reboosted into a higher orbit. This would
eliminate the need to launch a replacement for the
defective or decaying satellite.

CHARACTERISTICS:
- Physical Characteristics: Undetermined
- Potential For
  Technology
  Demonstration: Near-Term

CRITICAL ISSUES:
- Mechanisms and rendezvous techniques to capture satellite
- Compatibility with existing satellite systems
- Trade-off of the mission and reboost requirements

STATUS:
- Preliminary analysis indicates feasible concept
- No defined mission requirement

DISCUSSION: Integration of this system may be costly. The concept appears to be
feasible, but the practicality has not been established. No mission drivers have yet been
determined.

CONTACTS:
- Joe Carroll

REFERENCES:
G. Von Tiesenhausen, ed., Tether Applications Concept Sheets, June 28, 1984.

111

-- TRANSPORTATION --

Tether Rendezvous System

APPLICATION: Used to supplement the operations of
the Space Station and OMV.

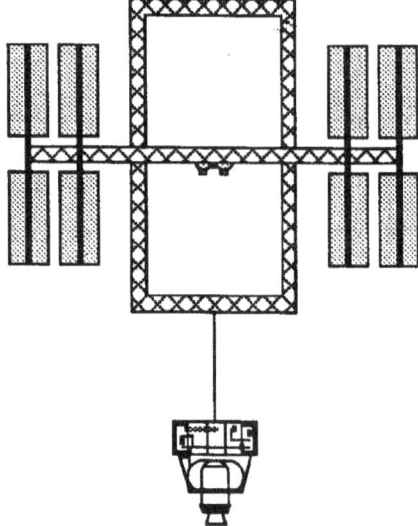

DESCRIPTION: The Tether Rendezvous System would be
used to capture and retrieve payloads, OTVs or the Space Shuttle
to the Space Station. The system would consist of a "smart"
hook which would be able to rendezvous and attach to a payload
with or without human intervention.

CHARACTERISTICS:
- Physical Characteristics: Undetermined
- Potential For
  Technology
  Demonstration: Mid-Term

CRITICAL ISSUES:
- Extent of system capabilities needs to be determined
- Dynamics in the tether and on the Space Station after rendezvous
- System design
- Rendezvous and capture techniques
- Hardware required

STATUS:
- Concept under study by Aeritalia
- Preliminary evaluations have been positive

DISCUSSION: The Tether Rendezvous System can supplement the operations of the
Space Station or any space platform by accomplishing remote rendezvous, increasing
flexibility, decreasing risk and saving a great amount of propellant for incoming vehicles
(STV, OMV, or the Shuttle Orbiter).

CONTACTS:
- Chris Rupp
- Joe Carroll
- Franco Bevilacqua

REFERENCES:
G. Von Tiesenhausen, ed., Tether Applications Concept Sheets, June 28, 1984.

Stuart, D. G., "Guidance and Control for Cooperative Tether-Mediated Orbital
Rendezvous," Journal of Spacecraft and Rockets, 1988.

-- TRANSPORTATION --

Upper Stage Boost from Orbiter

APPLICATION: Boost an upper stage payload into a higher orbit.

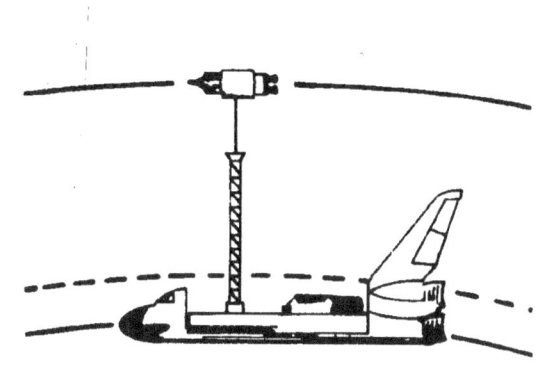

DESCRIPTION: An upper stage is deployed along a tether "upward" (away from the Earth) from the Shuttle Orbiter. Libration begins and momentum is transferred from the Shuttle to the upper stage, enhancing the performance envelope of the upper stage motor. A SEDS-derived (e.g. no retrieval capability) deployer system could be used. The Orbiter could be deboosted along with the upper stage boost. Spinup capability for some upper stages may be required.

CHARACTERISTICS:
* Length: Dependent on desired final orbit
* Tether Deployment
  System: Permanent or removable from Orbiter, TSS-derived
* Potential For
  Technology
  Demonstration: Near-Term

CRITICAL ISSUES:
* Requirement for spinup capability may be difficult

STATUS:
* Ball Brothers, Selected Tether Applications Study, Phase III
* SEDSAT project at University of Alabama in Huntsville
* SEDSAT deployment study at SAO

DISCUSSION: This application could be tailored to the Space Transfer Vehicle (STV). An expendable tether system or TSS-derived system could eliminate a major portion of the STV propellant required and increase payload capability for a specific mission with a fixed STV. The SEDSAT project (currently cancelled) was supposed to be the first space mission to boost a satellite into higher orbit with a tether . The boosting effect was observed at TSS-1R tether breakup

113

CONTACTS:
- James K. Harrison
- Les Johnson
- Enrico Lorenzini
- Mauro Pecchioli

REFERENCES:

"Study of Orbiting Constellations in Space," Contract RH4-394019, Martin Marietta, Smithsonian Astrophysical Observatory, December 1984.

Pecchioli, M., and Graziani, F., "A Thrusted Sling in Space: A Tether-assist Maneuver for Orbit Transfer," Second International Conference on Tethers In Space, Venice, Italy, 1987.

Proc. of Fourth International Conference on Tethers in Space, Washington DC, 10-14 April 1995

Applications "Satellite Boost from Orbiter" and "Small Expendable Deployer System"

Tether Assisted Transportation System (TATS)

APPLICATION: TATS is a tether-based system that provides the Space Station Alpha with transport capability not dependant on conventional propulsion

DESCRIPTION: The need and the feasibility of the additional Tether Assisted Transportation System have been evaluated in the context of the International Space Station Alpha. A preliminary cargo's traffic analysis indicated that large benefits in terms of mass and cost saving are expected by tether deorbit of disposable cargoes. The tether use was discovered to present also additional benefits increasing the safety of the Station and simplifying the execution of some operations.

CHARACTERISTICS:

- Mission Duration : up to some hours
- Altitude : 400-450 Km
- Active phase : < 1 day
- Return to ground : Re-entry Capsules
- Accommodation : Space Station
- Mass Deployer : 300 Kg (typical)
- Tether Mass : 40 Kg (typical))
- Capsule Mass: 150 Kg (typical)
- Tether length : about 37 Km

- Potential For Technology Demonstration: Near-Term

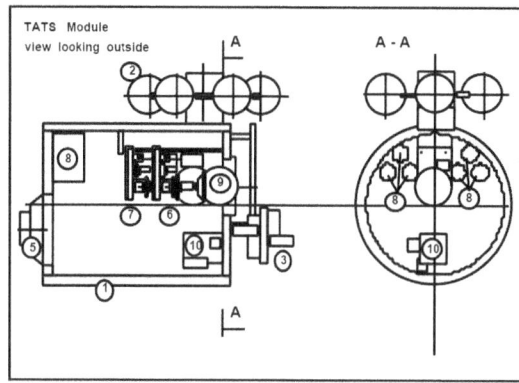

1 Structure of the TATS-module
2 Storage system for the re-entry capsules
3 Manipulator on rails
4 Air lock to the space
5 Docking mechanism and air lock to station
6 Tether system in operating position
7 Tether system in position during preparation
8 Storage system for replaceable tether units
9 Re-entry capsule in start position
10 Path tracking system

CRITICAL ISSUES:
- System configuration analysis,trade-off and design
- Re-entry capsule architecture definition
- Space Station-based Operations definition
- Station storage system for capsules and waste containers design
- Station robotic for TATS elements handling definition
- Tether system deployment timing for proper prograde swing
- Dynamics of tether after payload release

DISCUSSION: A potential utilization scenario of an additional Tether Assisted Transportation System has been devised to show the extent of its capabilities. As an example, the following evolution could be considered:

Initial Capability
- Frequent Sample Return
- Raduga-type Capsule Deorbit

Waste Disposal

Small Payloads Disposal

Full Capability
- Frequent Sample Return
- Raduga-type Capsule Deorbit
- Waste Disposal
- Cargoes Deorbit (PROGRESS, ATV)
- Large Modules and Payloads   Disposal

TATS consists of a set of re-entry capsules in a storage compartment, tools to allow the loading of the processed samples, a separation system (springs), and a tether deployer to perform properly capsule deployment and release. The analysis of possible ways to accommodate the TATS system on the Station has been focused on the two main options for accommodation: External and Internal Accommodation. Several possible options have been envisaged for possible accommodation of the system both at the ISSA US section and at the ISSA RS section.

CONTACTS:
- Pietro Merlina

REFERENCES:
"Tether Assisted Transportation System (TATS)", ESA/ESTEC contract No. 11439/95/NL/VK, Alenia Spazio/RSC Energia/DASA, 1995.

Failsafe Multiline Tethers for Long Tether Lifetimes (Hoytether)

APPLICATION: Long-life, damage resistant tether system for extended-duration, high-value, and crew-rated missions. Applications include low-drag, long life tethers for atmospheric and ionospheric science, electrodynamic tethers for in-orbit power and propulsion, and high-strength tethers for LEO-GEO-Lunar transport systems.

DESCRIPTION: The lifetimes of conventional single-line tethers are limited by damage due to meteorite and orbital debris impactors to periods on the order of weeks. Although single-line tether lifetimes can be improved by increasing the diameter of the tether, this incurs a prohibitive mass penalty. The Hoytether, shown in the figure, is a tether structure composed of multiple lines with redundant interlinking that is able to withstand many impacts.

Hoytether Section

CHARACTERISTICS:
- Can be designed to have survival probabilities of >99% for periods of months to years.

CRITICAL ISSUES:
Development of methods to fabricate and deploy many-kilometer long multiline tethers.

STATUS:
- 1/2 km long samples of bi- and tri- line Hoytethers were fabricated during a Phase I SBIR effort.
- A 1/2 km bi-line Hoytether was successfully deployed from a SEDS deployer ground tests.
- Development of methods for fabricating and deploying multi-kilometer conducting and non-conducting Hoytethers continues under a Phase II SBIR contract.

DISCUSSION: Analytical modeling, numerical simulation, and ground-based experimental testing of this design indicate that this tether structure can achieve lifetimes of tens of years without incurring a mass penalty. Moreover, while single-line tether survival probability drops exponentially with time, redundant linkage in failsafe multiline tethers keeps the tether survival probability very high until the tether lifetime is reached. The survival probability of a failsafe multiline tether is compared to that of an equal-mass single line tether in next figure.

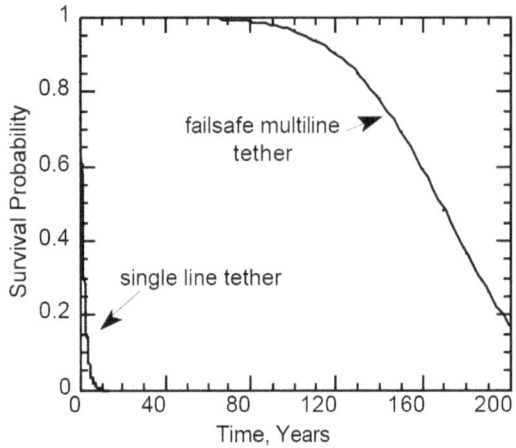

Lifetime comparison of equal-weight single line
and failsafe multiline tethers for a low-load mission.

CONTACTS:
- Robert P. Hoyt
- Robert L. Forward

REFERENCES:

Proceeding of the Fourth International Conference on Tethers in Space, Washington, DC, 10-14 April 1995.

R.L. Forward, R.P. Hoyt, Failsafe Multistrand Tether SEDS Technology Demonstration, Final Report on NAS8-40545 SBIR 94-1 Phase I Research Study.

R.L. Forward, Failsafe Multistrand Tethers for Space Propulsion, Forward Unlimited, Final Report on NAS8-39318 SBIR 91-1 Phase I Research Study.

SECTION 4.0 TETHER FUNDAMENTALS

4.1   GRAVITY GRADIENT

4.1.1   General

Gravity-gradient forces are fundamental to the general tether applications of controlled gravity, and the stabilization of tethered platforms and constellations. The basic physical principles behind gravity-gradient forces will be described in this section. This description will be in three parts. The first will discuss the principles behind the general concept of gravity-gradient forces. The second will continue the discussion, addressing the specific role of these forces in controlled-gravity applications. The third will address their role in the stabilization of tethered platforms and constellations.

For the purposes of this discussion, it will be sufficient to describe the motion of the simple "dumbbell" configuration, composed of two masses connected by a tether. Figure 4.1 shows the forces acting on this system at orbital velocity. When it is oriented such that there is a vertical separation between the two masses, the upper mass experiences a larger centrifugal than gravitational force, and the lower mass experiences a larger gravitational than centrifugal force. (The reason for this is described later in the discussion.) The result of this is a force couple applied to the system, forcing it into a vertical orientation. This orientation is stable with equal masses, and with unequal masses either above or below the center of gravity. Displacing the system from the local vertical produces restoring forces at each mass, which act to return the system to a vertical orientation. The restoring forces acting on the system are shown in Figure 4.2 (see Ref. 1).

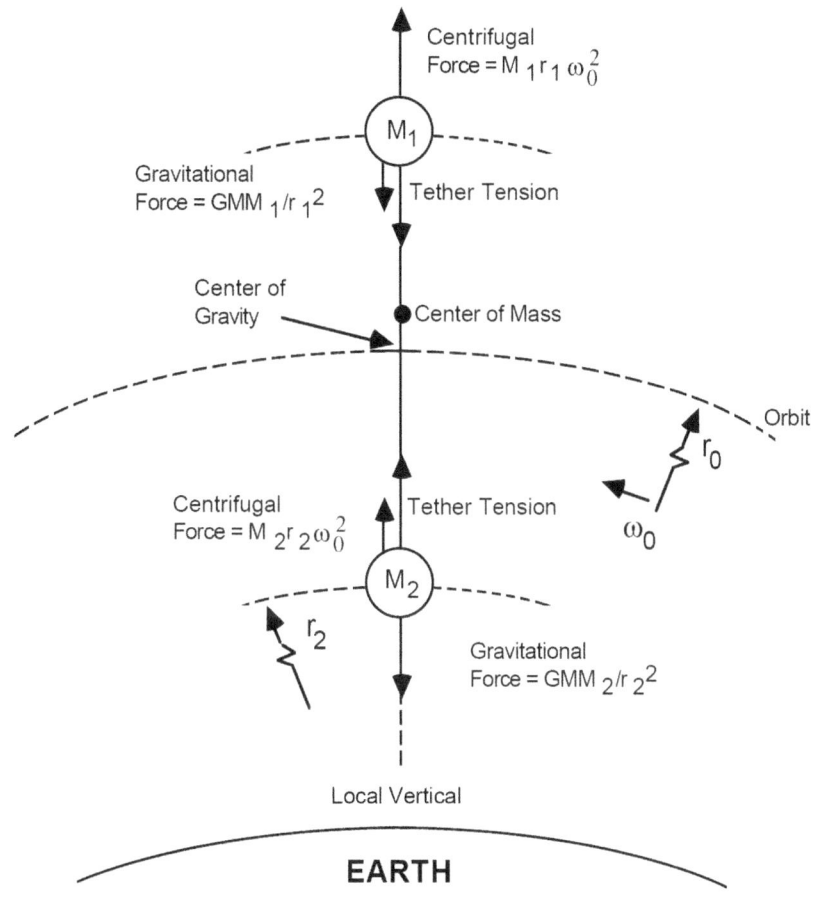

Figure 4.1 Forces on Tethered Satellites

Since the gravitational acceleration changes nonlinearly with distance from the center of the Earth, the center of gravity of the tethered system will not coincide exactly with its center of mass. The separation becomes more pronounced as the tether length increases. However, the separation is not dramatic for systems using less than very large long lengths. Therefore, for the purpose of this discussion it will be assumed that the center of mass coincides with the center of gravity. Furthermore, to facilitate an "uncluttered" discussion, the two masses will be assumed to be equal, and the tether mass will be ignored.

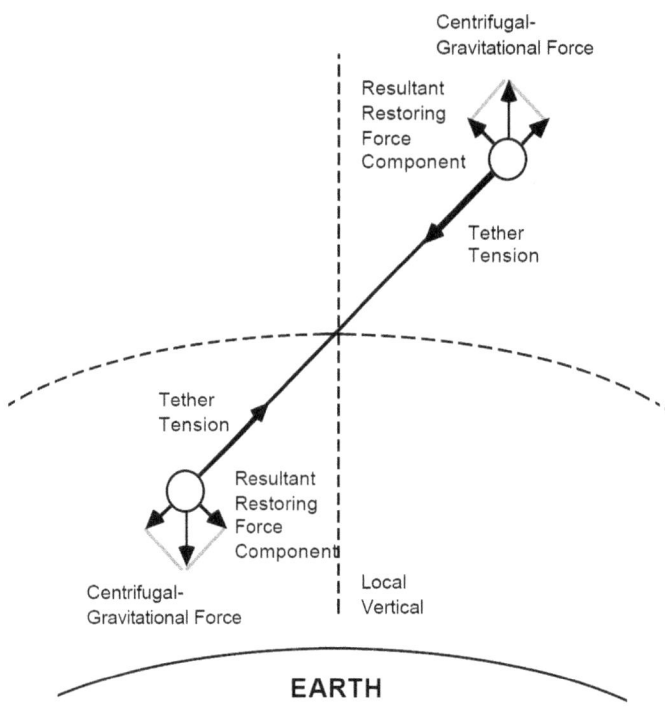

Figure 4.2 Restoring Forces on Tethered Satellites

The gravitational and centrifugal forces (accelerations) are equal and balanced at only one place: the system's center of gravity (C.G.). The center of gravity (or mass), located at the midpoint of the tether when the end masses are equal, is in free fall as it orbits the Earth, but the two end masses are not. They are constrained by the tether to orbit with the same angular velocity as the center of gravity. For the center of gravity in a Keplerian circular orbit, equating the gravitational and centrifugal force,

$$\frac{GMM_o}{r_o^2} = M_o r_o \omega_o^2 \qquad \text{and}$$

$$\omega_o^2 = \frac{GM}{r_o^3} \qquad ; \text{ where}$$

G = universal gravitational constant $(6.673 \times 10^{-11} \ Nm^2/kg^2)$,

M = mass of the Earth $(5.979 \times 10^{24} \ kg)$,

M_o = total tether system mass (kg),

r = radius of the system's center of gravity from the center of the Earth (m), and

ω_o = orbital angular velocity of the center of gravity (s^{-1}).

Since

122

$$\omega_o = \frac{V_o}{r_o} \qquad \text{and}$$

$$\omega_o = \frac{2\pi}{T_o} \qquad , \text{ where}$$

V_o = orbital speed of the center of gravity, (m/s), and
T_o = orbital period of the center of gravity (s),

$$V_o^2 = \frac{GM}{r_o} \qquad \text{and}$$

$$T_o^2 = \frac{4\pi^2 r_o^3}{GM}$$

Note that the orbital speed, period, and angular velocity depend on the orbital radius, and are independent of the tether system mass.

If the two end masses were in Keplerian circular orbits at their respective altitudes and were not connected by a tether, their orbital speeds would be different from the tethered configuration. For the upper mass, applying equations (1) and (2),

$$\omega_1^2 = \frac{GM}{(r_o + L)^3} \qquad \text{and}$$

$$V_1^2 = \frac{GM}{(r_o + L)} \qquad ; \text{ where}$$

L = tether length from the center of gravity to the mass (m).

Similarly, for the lower mass,

$$\omega_2^2 = \frac{GM}{(r_o - L)^3} \qquad \text{and}$$

$$V_2^2 = \frac{GM}{(r_o - L)}$$

It can be seen that without the tether, the upper mass would move at a slower speed and the lower mass would move at a higher speed. The tether, therefore, speeds up the upper mass and slows down the

123

lower mass. This is why the upper mass experiences a larger centrifugal than gravitational acceleration, and why the lower mass experiences a larger gravitational than centrifugal acceleration. The resulting upward acceleration of the upper mass and downward acceleration of the lower mass give rise to the balancing tether tension. They also produce the restoring forces when the system is deflected from a vertical orientation. The masses experience this tension as artificial gravity.

The artificial-gravity force and tether tension are equal to the gravity-gradient force. The gravity-gradient force on a mass, m, attached to the tether at a distance, L, from the system's center of gravity is equal to the difference between the centrifugal and gravitational forces on it. An approximate value for this force is given by,

$$F_{GG} \approx 3L \, m \, \omega_o^2$$

For mass m below the center of gravity, the gravity-gradient force is simply

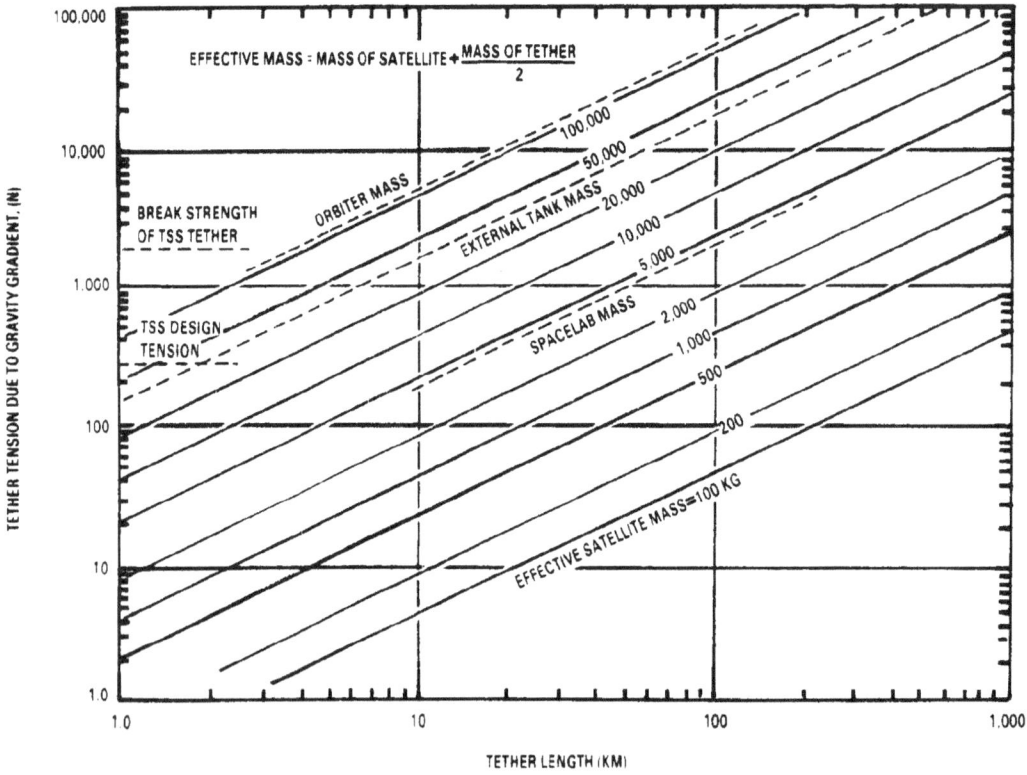

$$F_{GG} \approx -3L \, m \, \omega_o^2 \, ,$$

indicating that the gravity-gradient force acts upward above the center of gravity and downward below it. The force acts along the tether and away from the center of gravity. Furthermore, the gravity-gradient acceleration and force increase as the distance from the center of gravity increases and as the orbital radius

124

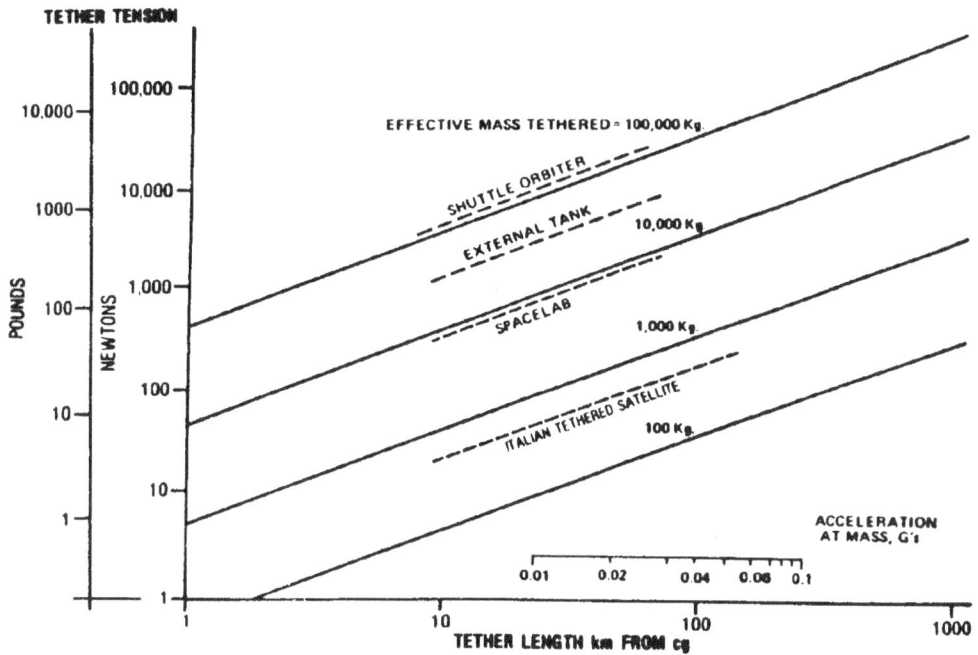

of the center of gravity decreases. (A more rigorous derivation of this equation is presented in Ref. 2, and also in Ref. 3). Figures 4.3 and 4.4 show the tether tension (artificial-gravity force) and artificial-gravity acceleration as a function of tether length from the center of gravity for various system masses in LEO (see Ref. 4). Figure 4.5 shows the tether mass and g-level as a function of tether length for a tether made of Kevlar 29. This figure includes tapered tethers which are discussed below.

Figure 4.3 Tether Tension Due to Gravity Gradient Versus Tether Length From
Center of Gravity and Effective Satellite Mass In LEO

Figure 4.4 "Artificial Gravity" at Tethered Masses in LEO

Figure 4.5 Tether Mass and g-Level Versus Tether Length for Kevlar 29 Tethers

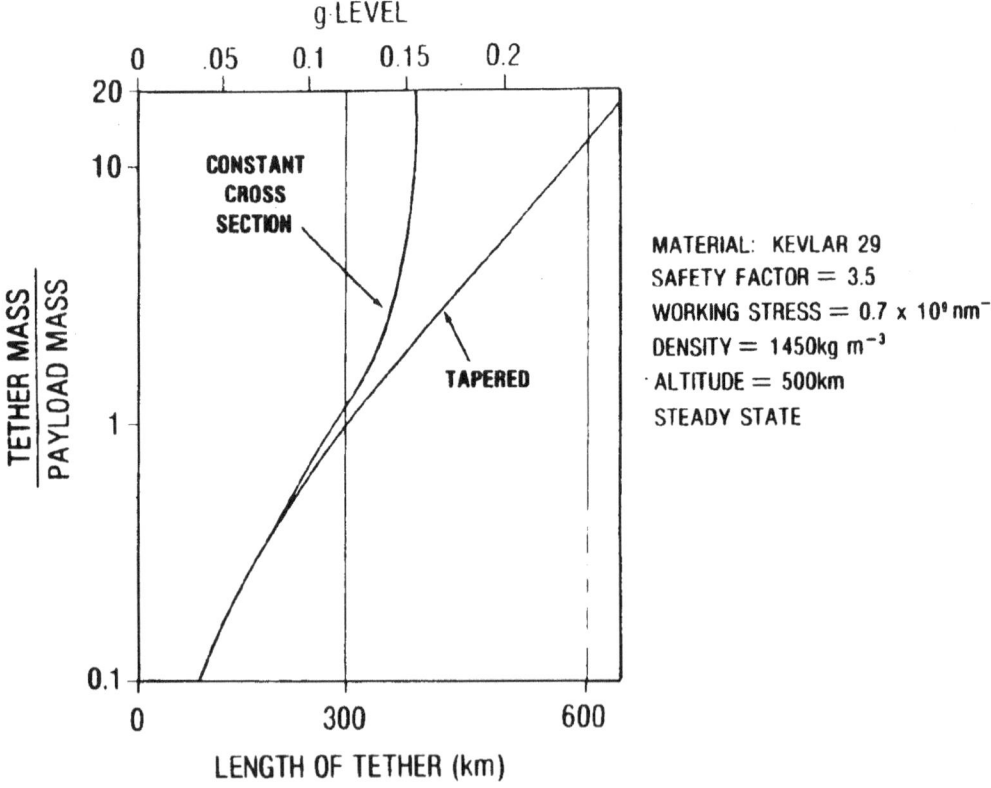

Since the gravity-gradient force and acceleration in orbit vary with GM/r_o^3 (where M is the planetary mass), they are independent of the planet's size, and linearly dependent on its density. The acceleration is largest around the inner planets and the Moon (0.3-0.4×10^{-3}g/km for low orbits, where g is Earth gravity), and about 60-80% less around the outer planets. The gravity-gradient acceleration decreases rapidly as the orbital radius increases (to 1.6×10^{-6} g/km in GEO).

Although the vertical orientation of the tether system is a stable one, there are forces which cause it to librate (oscillate) about the vertical. These weak but persistent forces include atmospheric drag due to the different air densities encountered in the northward and southward passes of non-equatorial orbits and due to solar heating and electrodynamic forces (for conducting tethers). Station-keeping and other rocket maneuvers would also contribute to driving (or damping) libration. The natural frequency for in-plane (in the orbit plane) librations is $\sqrt{3} \, \omega_o = 1.732 \, \omega_o$, and $2 \omega_o$ for out-of-plane librations (a detailed derivation is contained in Ref. 2).

Since both the displacement and restoring forces increase linearly with tether length, libration frequencies are independent of tether length. Therefore, the tether system will librate as a solid dumbbell (except for very long tethers, where the gravity gradient itself varies). Libration periods, however, do increase at large amplitudes. Since the tether constrains the motion of the masses, the sensed acceleration is always along the tether. Furthermore, the tether can go slack if the in-plane libration angle exceeds 65°, or if the out-of-plane libration angle exceeds 60°. The slackness can be overcome by reeling or unreeling the tether at an appropriate rate. Additional information on tether libration is presented in Ref. 5 and also Section 5.0.

Libration can be damped out by varying the tether length. It would be deployed when the tension was too high and retracted when the tension was too low. Since the in-plane and out-of-plane librations have different periods, they could be damped simultaneously. Shorter-period, higher-order tether vibrations could also be damped in this way.

Since the portion of the tether at the center of gravity must support the tether as well as the masses, the mass of long tethers must be taken into account. To minimize the tether's mass while maintaining its required strength, its cross-sectional area could be sized for a constant stress at all points along its length. The optimum design for very high tether tensions would be an exponentially tapered tether with a maximum area at the center of gravity and minima at the end masses. Tethers of constant cross-section have limited length, as indicated in Figure 4.5, whereas tapered tethers can have unlimited length; but then, its mass will increase exponentially along with its cross-section. A detailed discussion of tapered tether design is provided in Ref. 6.

In addition to the general areas of controlled gravity and tethered-platform and constellation stabilization, gravity-gradient effects play a fundamental role in applications related to momentum exchange and tethered-satellite deployment. These aspects are discussed in Section 4.3, entitled "Momentum Exchange."

127

4.1.2   Controlled Gravity

As a first step in discussing the role of gravity-gradient effects in controlled-gravity applications, a few definitions will be established. The definitions used in this book will be those recommended by the controlled gravity panel at the tether applications conference in Venice, Italy in October 1985 (Ref. 4). The term "controlled gravity" means the intentional establishment and control of the magnitude, vector properties, time dependence, and associated "noise" (uncertainty) of the acceleration field within a designated volume of space. In addition, the following definitions are also provided:

g = the acceleration on the equator at mean sea level on the Earth's surface (9.81 m/s^2);

microgravity = 10^{-4} g and smaller;

low gravity = 10^{-1} g to 10^{-4} g;

Earth gravity = 1 g;

hypergravity = greater than 1 g;

reduced gravity = microgravity and low gravity; and

enhanced gravity = hypergravity.

There are two basic tether configurations which can be used to provide controlled-acceleration fields: gravity-gradient-stabilized configurations (rotating once per orbit in an inertial frame), and rotating configurations (rotating more rapidly than once per orbit). This section will cover gravity-gradient-stabilized configurations. Rotating configurations are discussed later in Section 4.2.

In an orbiting, vertically-oriented, gravity-gradient-stabilized tether system composed of two end masses connected by a tether, all portions of each end mass experience the same acceleration, caused by the tether tension pulling on the end mass. This force is perceived as artificial gravity. As described before, its magnitude is proportional to the tether length from the system's center of gravity, and may be held constant or varied by deploying and retracting the tether. (For LEO, the gravity gradient is about 4 x 10^{-4} g/km.) Its direction is along the tether and away from the center of gravity.

This same principle can be used in more complex configurations (constellations) of three or more bodies. For example, consider a three-body system stabilized along the gravity gradient. In this system, a third body is attached to a crawler mechanism ("elevator") on the tether between the two primary end masses. The crawler mechanism allows the third body to be moved easily to any point along the tether between the end masses. The acceleration field (artificial gravity) in the third body can be controlled easily by moving it up or down the tether. Its distance from the system's center of gravity determines the magnitude of the artificial gravity within it. This artificial gravity acts in the direction along the tether and away from the center of gravity. The two end masses experience the artificial gravity determined by their distances from the center of gravity, as in the two-body system. The artificial gravity that they experience can also be held constant or varied by increasing or decreasing the tether length.

When positioned at the center of gravity, the third body could experience an acceleration field as low as about 10^{-8} g at the center of gravity, and 10^{-7} g and 10^{-6} g at distances from the center of gravity of

20 cm and 2 m, respectively. Using appropriate control laws, the third body's position could be automatically adjusted to produce a desired g-level time profile or to minimize transient disturbing effects.

Gravity-gradient effects can also be used to control the location of the system's center of gravity. This would be a very useful capability for the Space Station if microgravity experiments were to be performed on-board. Two tethered masses would be deployed vertically from the Space Station - one above and one below. By controlling the tether lengths, the position of the center of gravity could be maintained at a particular point in the system or moved to the other points as desired. This means that the artificial gravity at all points in the system would be correspondingly controlled to a fine degree of resolution. For example, the center of gravity could be adjusted to coincide with the minimum possible acceleration field.

All of these system configurations allow the generation and fine control of a wide range of g-levels. Using appropriate control laws, tether lengths and the relative positions of system components can be varied to produce desired gravity fields and their time profiles, to minimize transient disturbances to the gravity field, and to carefully control the location of the system's center of gravity. In addition to all of this, tethers also provide two-axis stabilization of the system.

Gravity-gradient systems have several advantages over rotating systems. They can provide artificial gravity for large-volume structures more easily. Also, the gravity gradient and Coriolis accelerations within these volumes are much less than those produced in rotating systems. One result of this is a lower occurrence of motion sickness. However, one disadvantage of gravity-gradient systems is that they would require very long tethers to achieve g-levels approaching 1 g or more. In fact, current tether materials are not strong enough to support their own weight at such tether lengths. However, by using moderate lengths and a relatively small rotation rate about the C.G, g-levels of 1 g or more can be achieved, with some increase in the Coriolis acceleration and gravity gradient. Figure 4.6 provides additional information concerning the acceptable values of artificial-gravity parameters (Ref. 4).

ARTIFICIAL GRAVITY
PARAMETERS

- UNAIDED TRACTION REQUIRES 0.1 G
- ANGULAR VELOCITY SHOULD BE LESS THAN 3.0 RPM TO AVOID MOTION SICKNESS
- MAXIMAL CENTRIPETAL ACCELERATION NEED NOT EXCEED EARTH GRAVITY
- CORIOLIS ACCELERATION SHOULD NOT EXCEED 0.25 CENTRIPETAL ACCELERATION FOR A LINEAR VELOCITY OF 3 FEET/SECOND IN A RADIAL DIRECTION
- GRADIENT SHOULD NOT EXCEED 0.01 G/FOOT IN RADIAL DIRECTION
- TETHER MASS MIGHT BE LIMITED TO 10,000 TO 20,000 POUNDS

ARTIFICIAL GRAVITY PARAMETERS

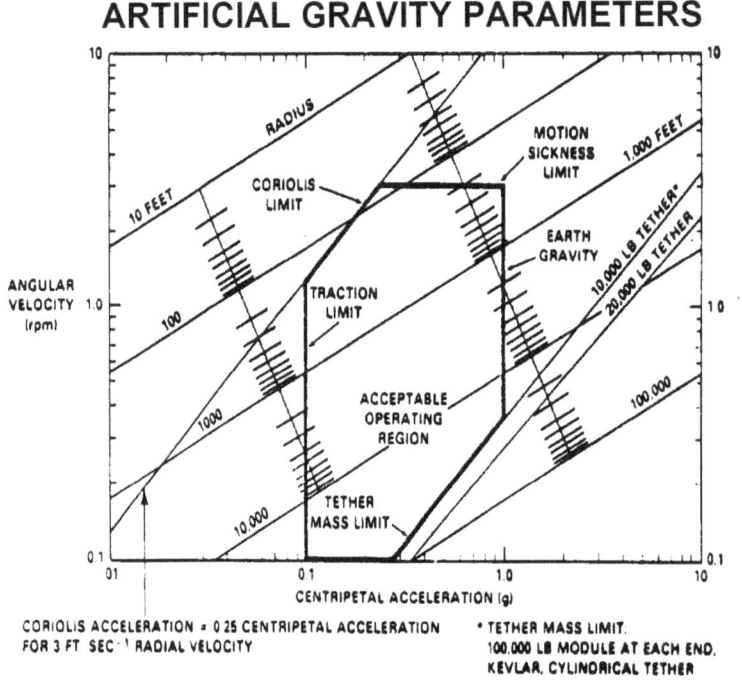

Figure 4.6 Acceptable Values of Artificial-Gravity Parameters

Tether technology suggests a number of exciting application possibilities. For example, since a tether can be used to attain a gravity field simply by deploying a counterweight along the gravity gradient, the establishment of a desirable low-level gravity on-board the Space Station appears practical. The use of 0.01 - 0.1 g on-board the Space Station might permit simpler and more reliable crew-support systems (such as eating aids, showers, toilets, etc.), operational advantages (no floating objects, easier tool usage, and panels and controls which are operated as in ground training), and perhaps some long-term biological advantages. The tether mass would be a significant part of the station mass to produce 0.1 g (using a

130

tapered 450 km tether), but would be relatively small for 0.05 g or less. However, careful consideration will have to be given to the disadvantages of tether system mass and complexity, and to assurance of survival in case of tether severing by meteoroid or debris impact. Such a system would also affect a microgravity laboratory, requiring it to be moved from the Space Station to the C.G. location.

A variable/low gravity laboratory module could be attached by a crawler mechanism to a tether deployed along the gravity gradient from the Space Station. A microgravity laboratory could also be built as part of the Space Station at its center of gravity. These labs could be used to examine the effects of microgravity and low gravity on both physical and biological processes. Some biological processes of interest would be plant and animal growth, and human performance and medical processes (such as those related to the cardiovascular, skeletal, and vestibular systems). The gravity-threshold values for various biological phenomena could also be studied. Such physical processes as crystal growth, fluid science, and chemical reactions could be studied. Many experiments in materials science and manufacturing could be performed in these gravity ranges. Liquid propellant storage and refueling facilities could be tethered to the Space Station. The artificial gravity produced by the tether would assist in propellant handling and transfer. Figure 4.7 shows the tether lengths necessary to allow propellant settling for the proper transfer of various propellants.

These are but a few of the possible applications of the artificial-gravity environments produced by gravity-gradient effects. Detailed descriptions of applications utilizing these gravity-gradient effects are contained in the "Tether Applications" (Section 3.0) of this handbook. Note that, due to the wide variety of possible system configurations, all of these applications are contained in one category. There are applications which overlap two or more categories and which could be logically listed under any one of them. In these cases, a judgment has been made as to which category is the most appropriate for the particular application and it is listed in that category. The applications related to the artificial gravity produced by gravity-gradient effects appears in the "Controlled Gravity" and "Space Station" categories of the "Applications" section, as appropriate.

Fluid Settling

- **SETTLING REQUIREMENT**
  - **- GRAVITY DOMINATE SURFACE TENSION**
- **FLUID SETTLING PARAMETER IS BOND NUMBER (B_o)**

$$B_o = \frac{\rho A D^2}{4\sigma}$$

ρ = FLUID DENSITY
σ = SURFACE TENSION COEFFICIENT
D = TANK DIAMETER

- **FLUID SETTLES IF $B_o > 10$**
  - **- $B_o = 50$ CHOSEN TO BE CONSERVATIVE**

PROPELLANT SETTLING ON A STATIC TETHER (Bo = 50)

Figure 4.7 Fluid Settling Properties of Various Liquid Propellants Under Conditions of Artificial Gravity - Required Tether Length Versus Propellant

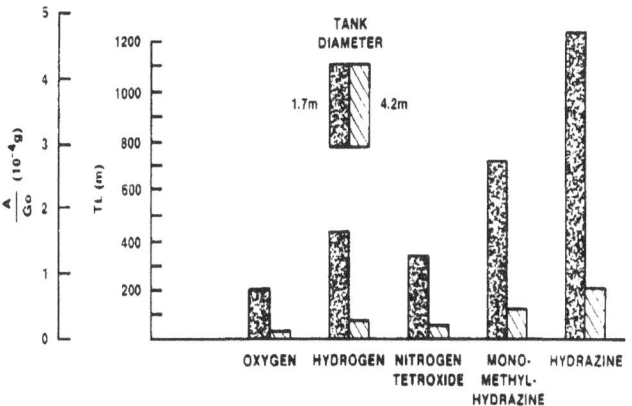

4.1.3 Constellations

Gravity-gradient forces also play a critical role in the stabilization of tethered constellations. A tethered constellation is defined as a generic distribution of more than two masses in space connected by tethers in a stable configuration. They can be configured in either one, two, or three dimensions. All of the non-negligible forces or gradients available in low orbit come into play to stabilize these various configurations. The vertical gravity gradient has the strongest influences, but differential air drag, electrodynamic forces, the J_{22} gravity component (an harmonic of the Earth's gravitational potential), and centrifugal forces also contribute. Different configurations utilize different combinations: 1-D vertical and horizontal, drag-and gravity-gradient-stabilized and electromagnetically stabilized (2-D).

Tethered constellations are divided into the two basic categories shown in Figure 4.8 (Ref. 4, p. 296). These are "static" and "dynamic" constellations. Static constellations are defined as constellations which do not rotate relative to the orbiting reference frame (they do rotate at the orbital rate when referred to an inertial frame). Dynamic constellations, on the other hand, are defined as constellations which do rotate with respect to the orbiting reference frame. These two basic categories are subdivided further. Static constellations include gravity-gradient-stabilized (one-dimensional, vertical), drag-stabilized (one- and gravity-gradient-stabilized (two-dimensional) constellations. Dynamic constellations include centrifugally stabilized two dimensional and three-dimensional constellations. This section will address only the static constellations.

Figure 4.8 Types of Tethered Constellations

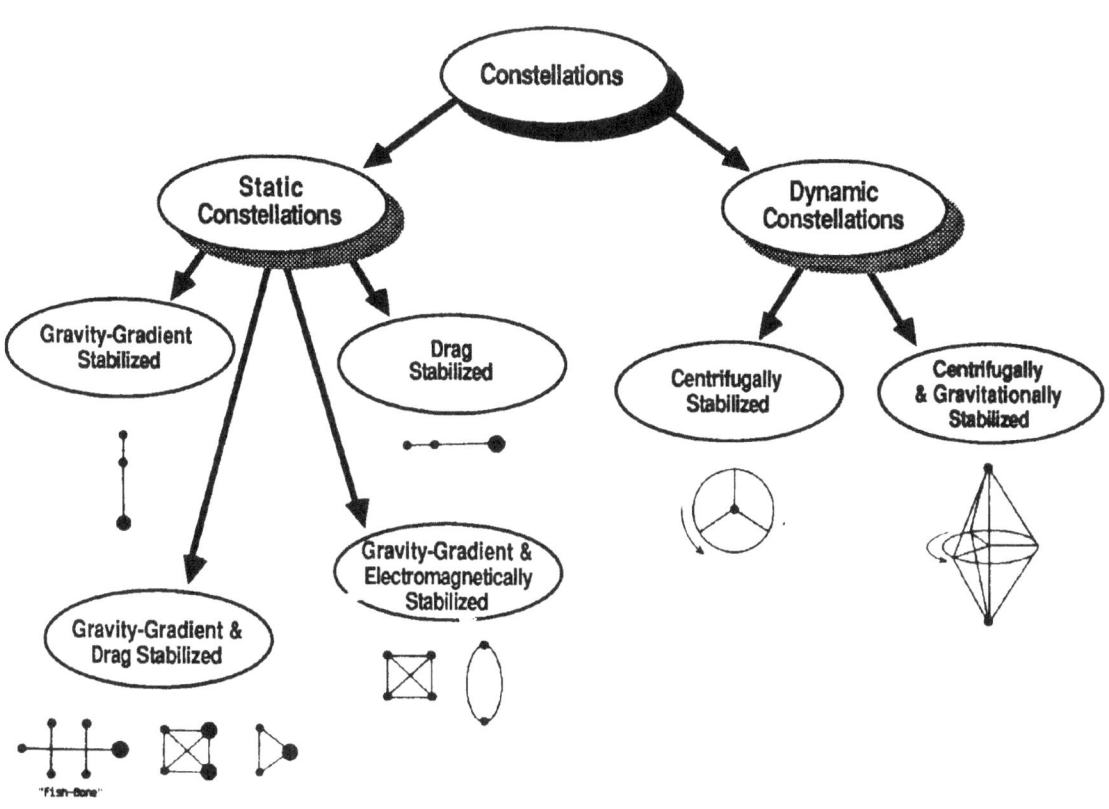

From the standpoint of stability and complexity, a gravity-gradient-stabilized, one-dimensional, vertical constellation is the most desirable configuration. A diagram showing three bodies tethered in this configuration is shown in Figure 4.9. Examples included the three-body configurations used for variable/low gravity and microgravity labs, and for the position control of the system center of gravity. Earlier discussion of vertical configurations included descriptions of their dynamics (including libration). The dominant influence on these constellations is the vertical gravity gradient.

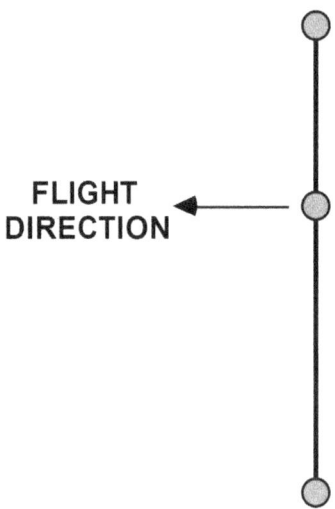

FLIGHT DIRECTION

Figure 4.9 Example Configuration of 1-D, Gravity-Gradient-Stabilized,
Vertical Constellation

Stability in one-dimensional, horizontal constellations is provided by tensioning the tethers. (Such a constellation is depicted in Figure 4.10.) By designing such a constellation so that the ballistic coefficient of each of its elements is lower than that of the element leading it and higher than that of the element trailing it, a tension is maintained in the tethers connecting them along the velocity vector. The resulting differential drag on its elements prevents the constellation from compressing, and the tension in its tethers prevents it from drifting apart. In principle, there is no limit to the number of platforms which can be connected in this manner. However, it should be noted that drag takes orbital energy out of the constellation, shortening its orbital lifetime unless compensated by some form of propulsion.

FLIGHT DIRECTION

Figure 4.10 Example Configuration of 1-D, Drag-Stabilized,
Horizontal Constellation

The fundamental parameter for one-dimensional, horizontal constellations is the differential ballistic coefficient of the two end bodies. In the case of a massive front body and a voluminous rear body (balloon), it is equal to the ballistic coefficient of the latter. Tether lengths and orbital lifetimes are competing requirements and are never sufficiently satisfied in the altitude range of interest. Since the vertical gravity gradient dominates over the differential air drag at the Space Station altitude and above, the maximum horizontal tether length must be short for stability. At lower altitudes (150-200 km) where the differential air drag becomes relatively strong, tether length may be longer, but the orbital lifetime will be limited.

The "fish-bone" configuration was the first proposed two-dimensional constellation and it utilizes both gravity-gradient and air-drag forces in order to attain its stability. A simple "fish-bone" constellation is depicted in Figure 4.11. For analytical purposes, this constellation can be reduced to an equivalent one-dimensional, horizontal constellation by lumping the overall ballistic coefficient of the rear leg (balloons plus tethers) and the front leg at the ends of the horizontal tether. Additional information on the stability analysis of the original "fish-bone" configuration shown in Figure 4.11 is presented in Ref. 4 (p.171-172) and contains calculated values of its stability limits versus altitude. Analysis has revealed that this configuration is less stable than a comparable one-dimensional, horizontal constellation. The necessity of a massive deployer at the center of the downstream vertical tether subsystem greatly reduces the area-to-mass ratio of that subsystem.

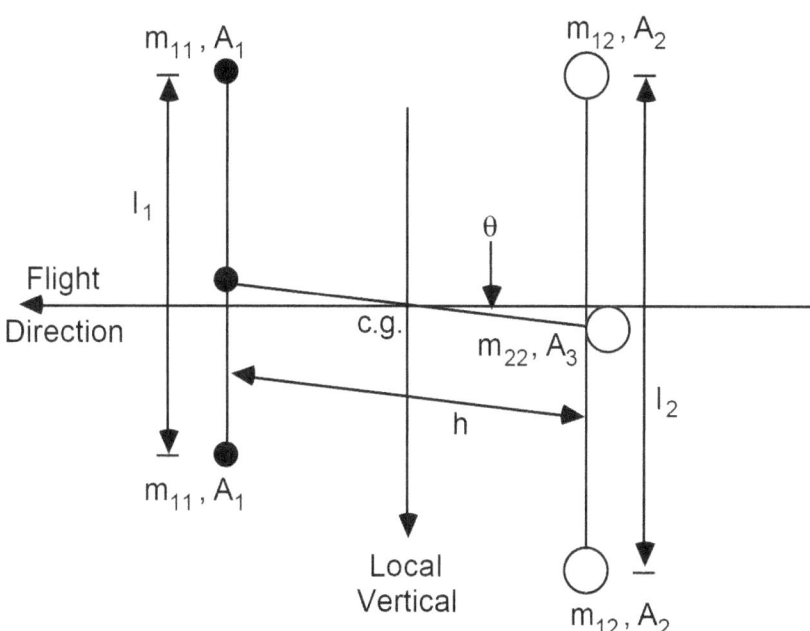

Figure 4.11 Example Configuration of 2-D, "Fish-Bone"

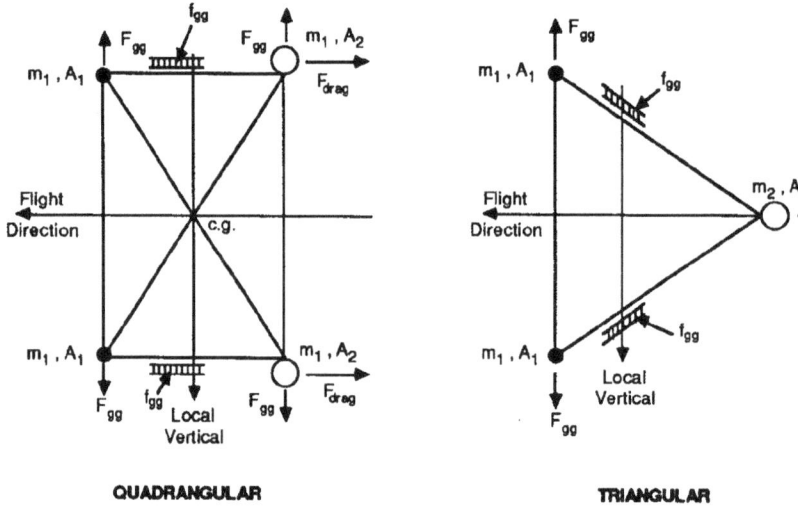

Two additional designs for a two-dimensional constellation, utilizing gravity-gradient and air-drag forces for stability, have been proposed. These drag-stabilized constellation (DSC) designs are depicted in Figure 4.12. With this type of configuration, the gravity gradient is exploited for overall attitude stability (the constellation's minimum axis of inertia must be along the local vertical), and differential air-drag forces are used to stretch the constellation horizontally for shape stability. The drag force is fully exploited to assure the minimum tension in the horizontal tethers, and not to counteract the gravity-gradient force as it does in the "fish-bone" configuration. Design parameters for DSC systems are presented in Ref. 4 (p. 175-178).

Two designs for a two-dimensional constellation utilizing gravity-gradient and electromagnetic forces for stability have been proposed. These electromagnetically stabilized constellation (ESC) designs are shown in Figure 4.13. In these configurations, the gravity gradient is again used for overall attitude stability (the minimum axis of inertia is vertical) and electromagnetic forces are used to stretch the constellation horizontally for shape stability. (These electromagnetic forces are discussed in detail in Ref. 7 and section 4.4).

Figure 4.12 Two Designs of 2-D DSC Constellations Horizontally

In the quadrangular configuration, current flows in the outer-loop tethers, interacting with the Earth's magnetic field, to generate electromagnetic forces in the outer loop. The current direction is chosen such that these forces push the tethers outward, tensioning them (like air inside a balloon). Although the shape is different in the pseudo-elliptical constellation (PEC) design, the same principle of electromagnetic tensioning of the outer-loop tethers is applied. The two lumped masses provide extra attitude stability without affecting the constellation shape. Moreover, since the resultant force is zero, the orbital decay rate is provided by air drag only. Design parameters for ESC systems are presented in Ref. 4 (p. 176-177).

138

Figure 4.13 Two Designs of ESC 2-D Constellations Where Shape Stability is
Provided by Electromagnetic forces

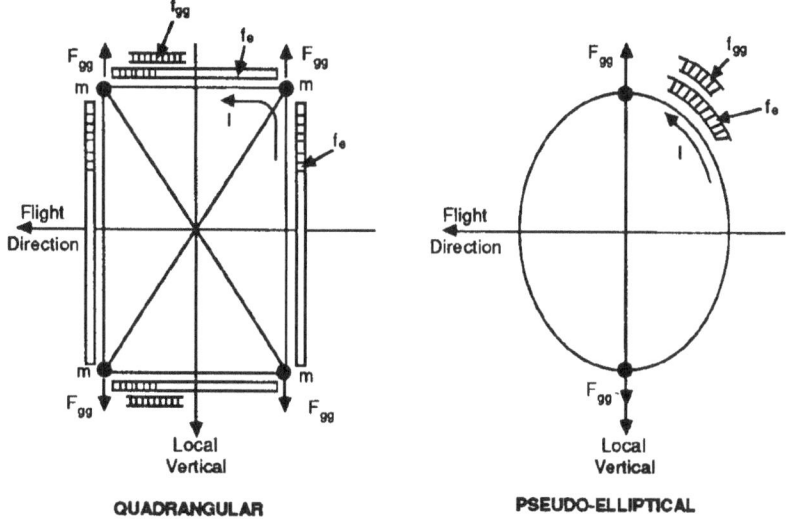

Preliminary conclusions on the design of two-dimensional constellations have been reached. The "fish-bone" constellations are less stable than the one-dimensional, horizontal constellations. "Fish-bone" constellations are stable with very short horizontal tethers (less than 100 m at 500 km altitude). The alternative quadrangular DSC and ESC constellations (and PECs for special applications) exhibit a better static stability. Suitable design parameters can provide good stability with a reasonably low power requirement for ESCs and feasible balloons for DSCs.

Typical dimensions for these constellations are 10 km (horizontal) by 20 km (vertical) with balloon diameters of about 100 m for DSCs, a power consumption of about 5.5 kW for ESCs and 2 kW for PECs. The ESC constellations have greater tension in the horizontal tethers than the DSC constellations and an orbital decay which is smaller by an order of magnitude. ESCs are suitable for low inclination orbits. Moreover, since they tend to orient their longitudinal plane perpendicular to the Earth's magnetic field (B vector), a small oscillation about the vertical axis at the orbital frequency is unavoidable even at low orbital inclinations. DSCs, on the other hand, are suitable for any orbital inclination. In the DSCs, the yaw oscillation occurs at high inclinations only due to the Earth's rotating atmosphere.

There are several proposed applications for one-dimensional, vertical constellations. A three-body configuration could be used for microgravity/variable-gravity laboratories attached to the Space Station or the Shuttle. A three-body system could be used on the Space Station to control the location of the center of gravity. A system of 3 or more bodies attached to the Shuttle or Space Station could be used as a multi-probe lab for the measurement of the gradients of geophysical quantities. A 3-body system could also function as an ELF/ULF antenna by allowing a current to flow alternatively in the upper and lower tether to inject an electromagnetic wave with a square waveform into the ionosphere. A space elevator (or crawler) for the Space Station is yet another application.

There are several proposed applications for two-dimensional constellations. An electromagnetically stabilized constellation could provide an external stable frame for giant orbiting reflectors. Multi-mass constellations in general allow a separation of different activities while keeping them physically connected, such as for power distribution, etc. Detailed analysis of these two-dimensional structures may be found in Ref. 7.

4.2   ROTATION OF TETHER SYSTEMS

4.2.1   General

Tethers will almost always be involved in some form of rotational configuration. Any planet-orbiting tether system, by nature, will rotate about the planet at the orbit angular velocity. The combination of the centrifugal forces due to rotation and gravity gradient acting on the tether end masses causes it to be stabilized in a vertical position about the planet center of mass. In many interplanetary applications, rotation will be desired to cause an artificial-gravity environment or to create a centrifugally stabilized configuration.

4.2.2 Controlled Gravity

A tether-mass system may desire controlled gravity for a number of applications. These may range from an artificial-gravity environment for manned interplanetary missions to a controlled-gravity platform for industrial space applications. The calculation of the acceleration at a point for purely circular motion is presented here. With reference to Figure 4.14, we assume that point P (which would represent the mass) is at a constant radius, r (the tether), from the center of our rotation system.

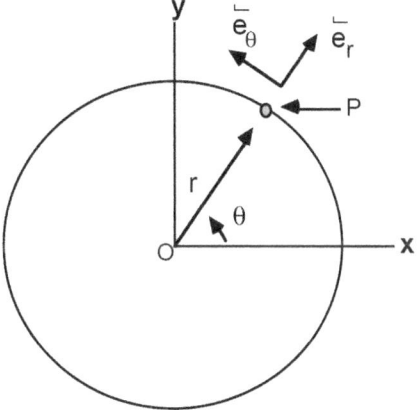

Figure 4.14 Circular Motion of a Point.

The acceleration can then be found by the expression:

$$\vec{a} = (-r\,\omega^2)\,\vec{e}_r + (r\,\dot{\omega})\,\vec{e}_\theta \quad ;$$

where,

$\vec{a}$ = acceleration at the point P (m/s^2),

$\vec{e}_r$ = unit vector in radial direction,

$\vec{e}_\theta$ = unit vector in tangential (velocity) direction,

r = radius (length of tether) (m),

ω = angular velocity (rad/s),

$\dot{\omega}$ = angular acceleration (rad/s^2).

Notice that if the angular velocity is constant the acceleration simplifies to

$$\vec{a} = (-r\,\omega^2)\,\vec{e}_r \quad ;$$

141

where the negative sign indicates that the acceleration acts toward the center of rotation (see Ref. 8).

As an example, suppose it is desired to calculate the gravity level at a manned module rotating about another similar module with angular velocity of 2.0 rpm, attached by a tether of length 200 meters. The center of mass will be exactly between them, and, with this as the origin, the distance to each module is 100 meters. Then, the calculation is,

$$a = r\omega^2$$

$$= (100 \text{ m}) \left[\left(\frac{2 \text{ rev}}{\text{min}} \right) \left(\frac{1 \text{ min}}{60 \text{ sec}} \right) \left(\frac{2\pi \text{ rad}}{\text{rev}} \right) \right]^2$$

$$= 4.38 \text{ m/s}^2$$

To calculate the gravity level (as compared to Earth's):

$$a = \frac{4.38 \text{ m/s}^2}{9.8 \text{ m/s}^2}$$

$$= 0.45 \text{ g} .$$

4.3 MOMENTUM EXCHANGE

4.3.1 General-Conservation of Angular Momentum

Tethers can have useful space applications by redistributing the orbital angular momentum of a system. A tether can neither create nor destroy system angular momentum, only transfer it from one body to another. Angular momentum is defined (for a rotating system, Figure 4.15) as,

$$\vec{h} = m\vec{r} \times \vec{v} = mr^2\vec{\omega} ;$$

where

$\vec{h}$ = angular momentum of system ($kgm^2 s^{-1}$),

m = mass of system (kg)

$\vec{r}$ = radius vector from center of rotating coordinate system (usually the Earth) to system center of mass (m),

$\vec{v}$ = velocity of system center of mass normal to r (ms^{-1}), and

$\vec{\omega}$ = system angular velocity (s^{-1}).

143

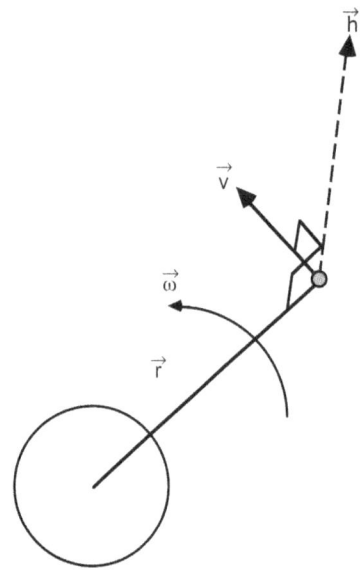

In genera ... ions using different
momentum excha ... by examples of their
application. A us ... ata".

4.3.2 Tether P:

Consider : ... ier as in Figure 4.16
(see Ref. 9).

In order to ... downward from the
Shuttle (M_1), it is ... t separation. After a
certain length of ... bits so that gravity-
gradient and cent ... nstrained by a tether,
mass M_1 would a ... igher orbital circular
velocity in ther ne ... s's gravitational field,

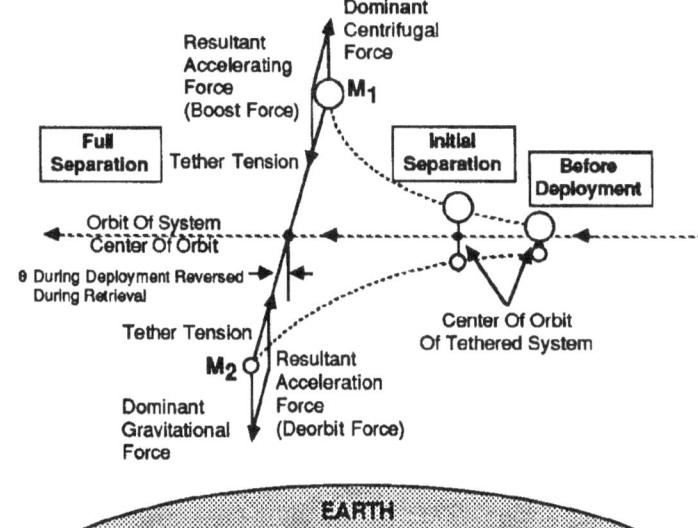

its potential energy is raised and its kinetic energy is lowered. For M_2 the exact opposite is true. Since the masses are constrained by a tether, they also must move at the same orbital velocity. Mass M_2, therefore, will "drag" mass M_1 along until libration occurs. Libration (pendulum motion) will continue due to the centrifugal, gravitational, and tether tension restoring forces.

144

Figure 4.16 Tethered Deployment

In this case, mass M_1 gained angular momentum equal to an identical amount lost by M_2. This amount of angular momentum transferred is equal to:

$$\Delta h = M_1 \ V \ \Delta R_1 = M_2 \ V \ \Delta R_2 \ .$$

The momentum is transferred from M_1 to M_2 through the horizontal component of the tether tension. This tension is caused by the Coriolis term of the acceleration expression of the librating masses.

If the tether is now cut, the upper mass, M_1, is boosted into an elliptical orbit having higher energy than it would have had due to its greater velocity. The point in the orbit where the tether is severed will correspond to the perigee of M_1. The situation is exactly reversed for M_2, which will be at its apogee at this point.

The preceding discussion explains the basic mechanics of momentum transfer in tethers. There are many variations of tethered deployment, many of which are beyond the scope of this text. Only some of the more basic ones will be described here.

Static and dynamic tether deployment are basically the same, except that static deployment occurs with the tether remaining under small angular displacements from the vertical, and dynamic deployments utilize large angular displacements. For certain dynamic deployments, it is possible to impart additional energy to one mass at the expense of the other. In order to implement this exchange, the deployment begins with a large angular displacement, tether tension is purposely kept low until a desired length is reached. When brakes are applied, a large angle prograde swing occurs. When the upper mass (payload) leads the lower mass, the tether is severed. In this way, an added boost due to the additional velocity of the prograde swing is accomplished.

Another method of tethered deployment is libration pumping. The tether is initially deployed then alternately extended and retrieved in resonance with tether tension variations during libration. (In-plane libration causes these tension variations due to Coriolis effects.) Spin pumping is yet another method, whereby libration pumping is carried further to the point that the tether system is caused to spin. In both cases, the added energy increases the departure velocity of the payload, just as in the dynamic tethered deployment case.

4.3.3   Orbit Variations

If the payload deployment described previously is carefully done, the orbits of both masses can be changed for one or both of their benefits. The Shuttle, for example, can boost a payload into a higher orbit and at the same time deboost itself back to Earth. Conversely, the Shuttle could perform a tethered deployment of its external tanks, whereby the tanks are deboosted back to Earth and the Shuttle is boosted to a higher orbit. Applications such as these are termed "momentum scavenging" since excess momentum is utilized for a beneficial purpose. The trick with this approach is that excess momentum must be available. One major application which is described in the applications section of the handbook is the Space Station-Shuttle deboost operation. This is an excellent example where both masses benefit. Resupply missions of the Space Station by the Shuttle are finalized by a tethered deployment of the Shuttle. In this way, the Space Station is boosted to a higher orbit and the Shuttle is de-boosted back to Earth. In order to utilize the additional momentum of the Space Station, tethered deployments of an STV are alternated with those of the Shuttle. Fuel savings can be obtained by both Shuttle and STV in this example. Tethers can also be used to change orbit eccentricity. This is done by libration pumping of tethered mass, phased as in Figure 4.17 (Ref. 9).

146

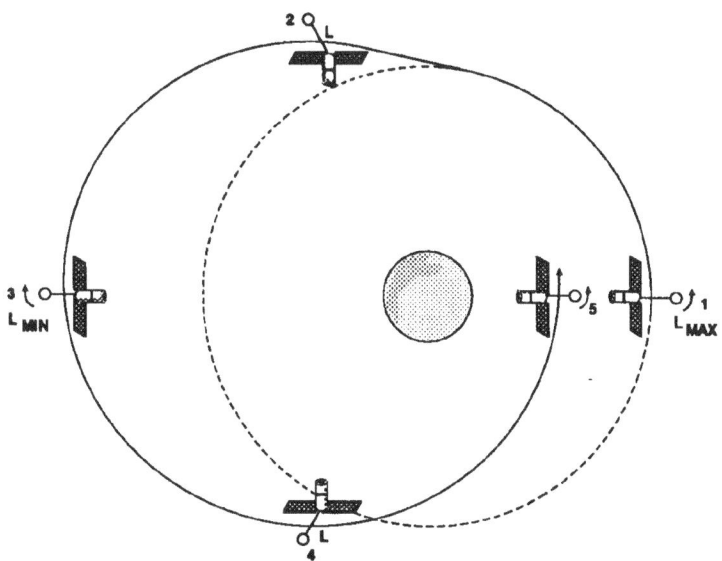

Figure 4.17 Orbit Eccentricity Change

147

At (1) the mass is fully extended, and libration commences. At (2), with the mass in a prograde swing, the retrieval motor pulls the spacecraft toward the mass, adding energy to the orbit (through the use of excess electrical energy transferred to the motor). At (3), which is the new apogee of the orbit, the tether length is at a minimum. At (4), with the mass in a retrograde swing, the tether is re-deployed and the retrieval brakes are used to dissipate orbital energy in the form of excess heat. At (5), the new perigee, the mass is again fully deployed.

4.4   ELECTRODYNAMICS

4.4.1   General

Electrodynamic tether systems can be designed to produce several useful effects by interacting with magnetic fields. They can be designed to produce either electrical power or thrust (either a propulsive thrust or a drag). They can also be designed to alternately produce electrical power and thrust. In addition, they can be designed to produce ULF/ELF/VLF electromagnetic signals in the upper atmosphere, and shape-stability for orbiting satellite constellations. Electrodynamic systems can be designed to produce electrical power.

4.4.2   Electric Power Generators

The discussion of electric power generation by tether systems will begin with electrodynamic systems in low Earth orbit. Consider a vertical, gravity-gradient-stabilized, insulated, conducting tether, which is terminated at both ends by plasma contactors. A typical configuration is shown in Figure 4.18 (Ref. 9, 10). As this system orbits the Earth, it cuts across the geomagnetic field from west to east at about 8 km/s. An electromotive force (emf) is induced across the length of the tether. This emf is given by the equation:

$$V = \int (\vec{v} \times \vec{B}) \cdot \vec{dl}$$
$$\text{along length of tether}$$

where

V = induced emf across the tether length (volts),

$\vec{v}$ = tether velocity relative to the geomagnetic field (m/s)

$\vec{B}$ = magnetic field strength (webers/m^2), and

$\vec{dl}$ = differential element of tether length - a vector pointing in the direction of positive current flow (m).

For the special case where the tether is straight and perpendicular to the magnetic field lines everywhere along its length, the equation for the emf simplifies to:

$$V = (\vec{v} \times \vec{B}) \cdot \vec{L} \quad ;$$

where

$\vec{L}$ = tether length - a vector pointing in the direction of positive current flow (m).

The equation for the induced emf across the tether in this special case can also be written as:

$$V = L v B \sin\theta \quad ;$$

where

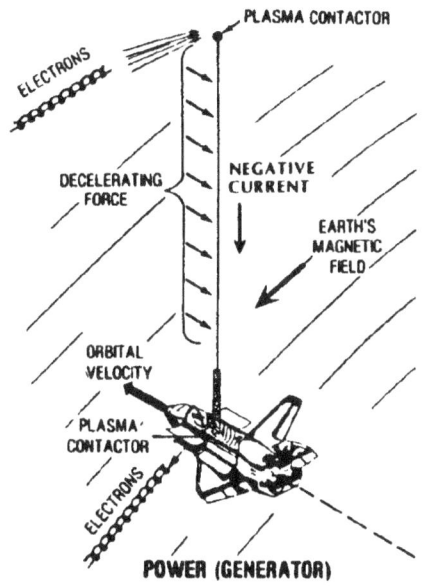

POWER (GENERATOR)

θ = angle between

(From these equations, it can be seen that equatorial and low-inclination orbits will produce the largest emfs, since the maximum emf is produced when the tether velocity and the magnetic field are perpendicular to each other.)

149

Figure 4.18 Power Generation With an Electrodynamic Tether

The emf acts to create a potential difference across the tether by making the upper end of the tether positive with respect to the lower end. In order to produce a current from this potential difference, the tether ends must make electrical contact with the Earth's plasma environment. Plasma contactors at the tether ends provide this contact, establishing a current loop (a so-called "phantom loop") through the tether, external plasma, and ionosphere. Although processes in the plasma and ionosphere are not clearly understood at this time, it is believed that the current path is like that shown in Figure 4.19. The collection of electrons from the plasma at the top end of the tether and their emission from the bottom end creates a net-positive charge cloud (or region) at the top end, and a net-negative charge cloud at the bottom. The excess free charges are constrained to move along the geomagnetic field lines intercepted by the tether ends until they reach the vicinity of the E region of the lower ionosphere where there are sufficient collisions with neutral particles to allow the electrons to migrate across the field lines and complete the circuit.

To optimize the ionosphere's ability to sustain a tether current, the tether current density at each end must not exceed the external ionospheric current density. Plasma contactors must effectively spread the tether current over a large enough area to reduce the current densities to the necessary levels. Three basic tether system configurations, using three types of plasma contactors, have been considered. They are: (1) a passive large-area conductor at both tether ends; (2) a passive large-area conductor at the upper end and an electron gun at the lower end; and, (3) a plasma-generating hollow cathode at both ends.

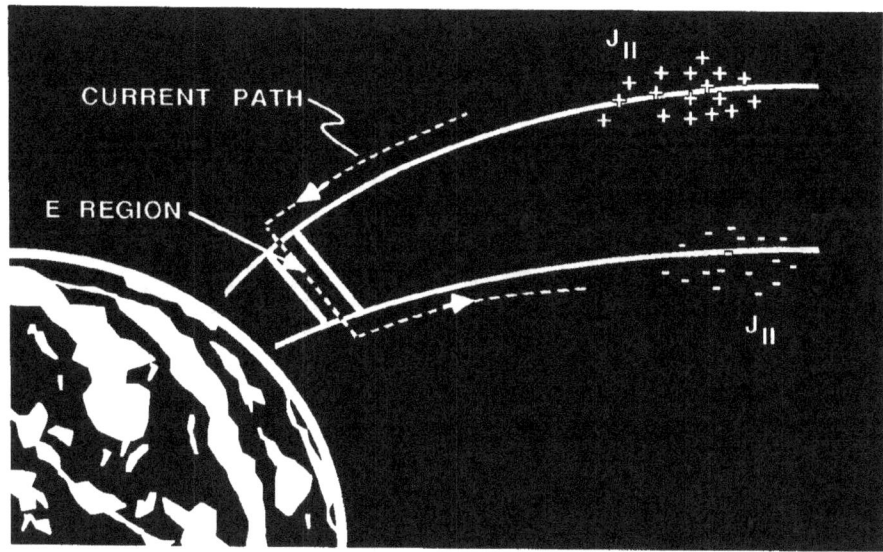

Figure 4.19 The Current Path External To An Orbiting
Electrodynamic Tether System

In the first configuration, the upper conductor (probably a conducting balloon) collects electrons. The lower plasma contactor in this configuration (perhaps a conductive surface of the attached spacecraft) utilizes its large surface area in a similar way to collect ions.

To achieve higher currents, it is possible to replace the passive large-area conductor at the lower end with an electron gun, providing the equivalent of collecting a positive ion current by ejecting a negative electron current. Ejecting these electrons at a high energy distributes them over an effectively large contact region. Unfortunately, electron guns are active plasma contactors, requiring on-board electrical power to drive them.

The third configuration is quite different from the first two. Based upon research results and performance modeling up to this point, it is projected to be the most promising of the three systems. Instead of relying on a passive and physically large conducting surface to collect currents, a hollow cathode at each tether end generates an expanding cloud of highly conductive plasma. The plasma density is very high at the tip of the tether and falls off to ionospheric densities at a large distance from the tip. This plasma provides a sufficient thermal electron density to carry the full tether current in either direction at any distance from the tether end, until it is merged into the ambient ionospheric plasma currents. This case of current reversibility allows the system to function alternately as either a generator or a thruster, with greater ease than either of the other two configurations (as will be discussed in more detail in the next section). Hollow cathodes are also active plasma contactors, requiring on-board electrical power and a gas supply to operate. However, they require much less power than an electron gun, and the gas supply should not impose a severe weight penalty. Two diagrams of a hollow cathode plasma source are shown in Figure 4.20. Additional diagrams and information relating to the construction and operation of the PMG hollow cathode plasma contactor are given in Figures 4.21, 4.22 and 4.23. Typical characteristics of a hollow cathode and an electron gun are compared in Figures 4.24 and 4.25. More information on the TSS and PMG flights results are given in Section 1.

151

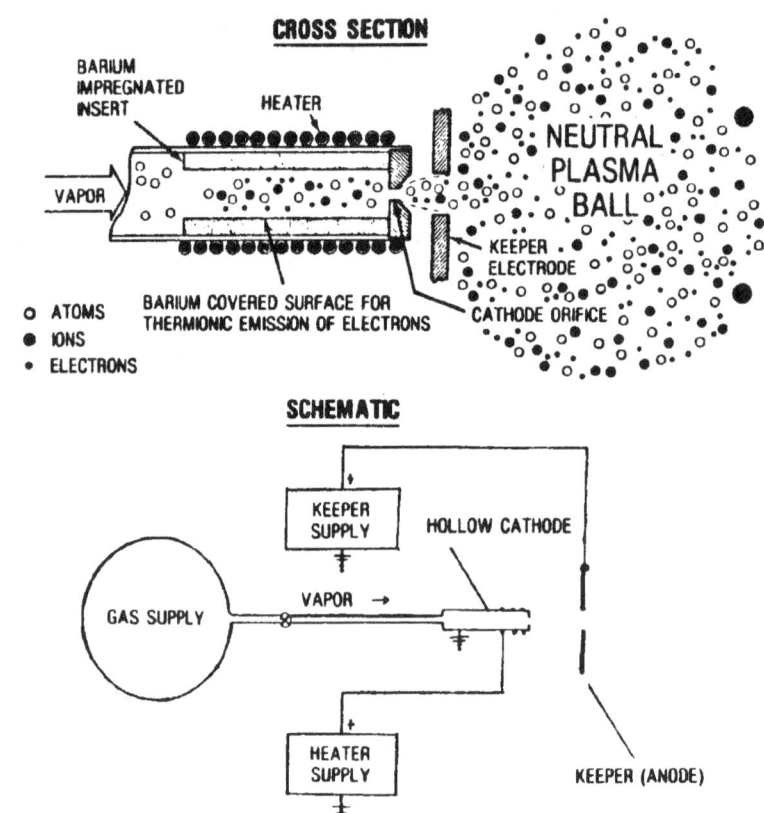

CROSS SECTION

BARIUM IMPREGNATED INSERT

HEATER

NEUTRAL PLASMA BALL

KEEPER ELECTRODE

CATHODE ORIFICE

VAPOR

o ATOMS
● IONS
· ELECTRONS

BARIUM COVERED SURFACE FOR THERMIONIC EMISSION OF ELECTRONS

SCHEMATIC

KEEPER SUPPLY

HOLLOW CATHODE

GAS SUPPLY

VAPOR →

HEATER SUPPLY

KEEPER (ANODE)

TSS-1R flight showed that larger tether currents can be generated and at much lower satellite potentials than were theoretically predicted. A final assessment on the performance of hollow cathodes flown on PMG compared to other configurations cannot be given as long as TSS-1R data analysis is completed and more flights test are verified. In addition, there may be particular applications for which passive contactors or electron guns are desirable. On the other hand, using hollow cathode plasma contactors should also be safer for spacecraft systems, since they establish a known vehicle ground reference potential with respect to the local plasma.

Figure 4.20 Diagrams of a Hollow Cathode Plasma Contactor

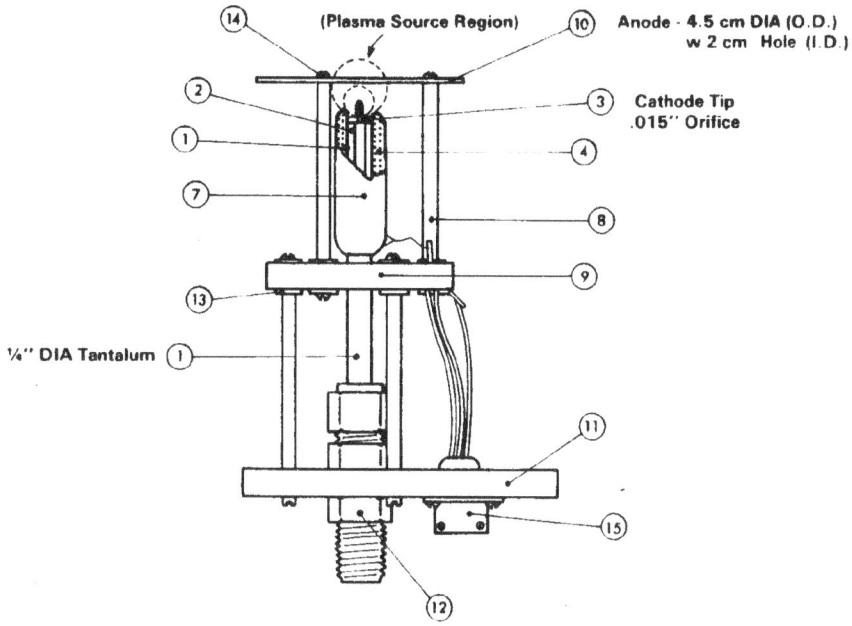

Figure 4.21 Diagram of the Plasma Motor/Generator (PMG) Hollow Cathode Assembly

153

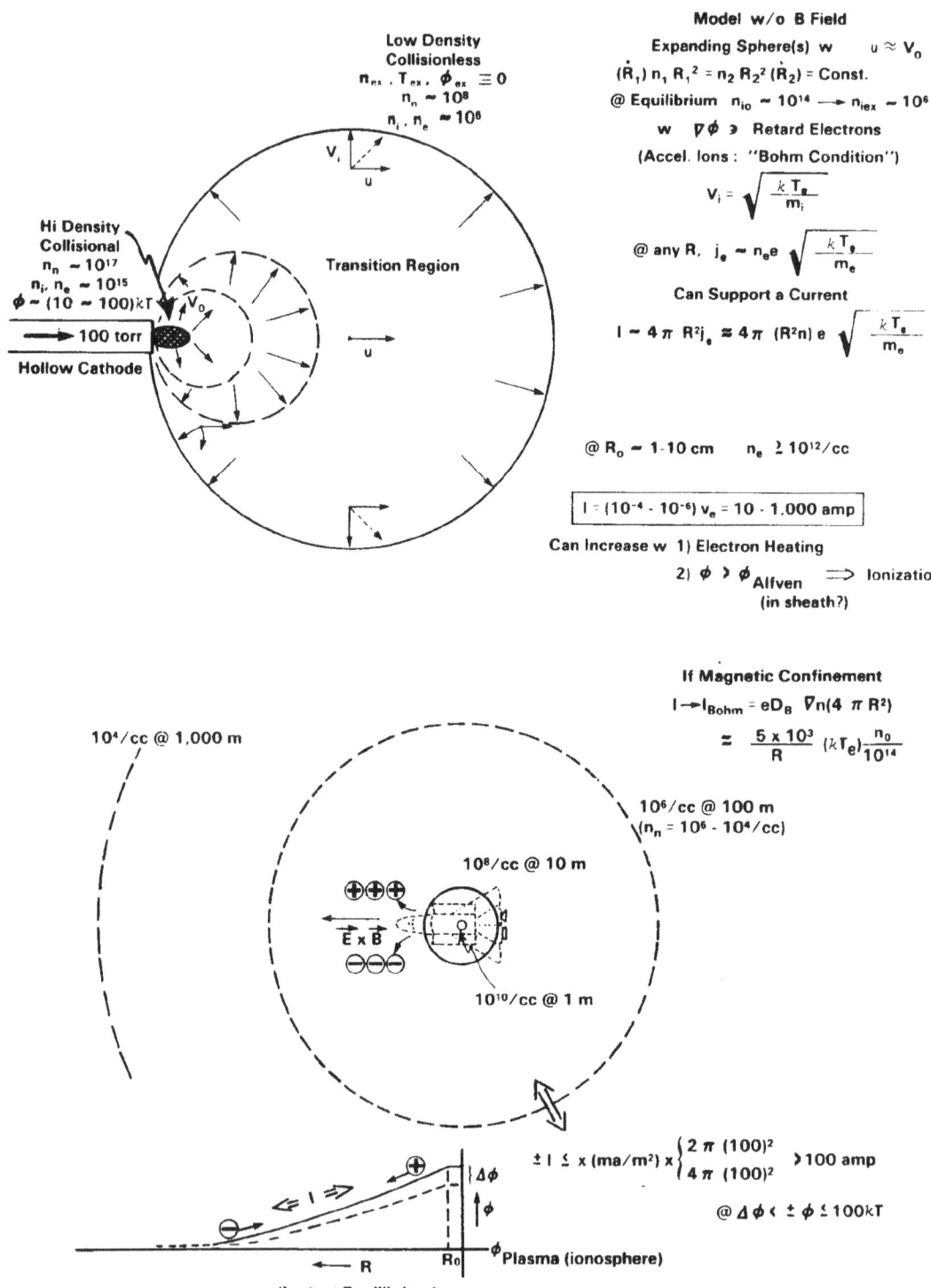

Figure 4.22 Plasma Cloud Expansion for PMG Hollow Cathode Plasma Contactor

154

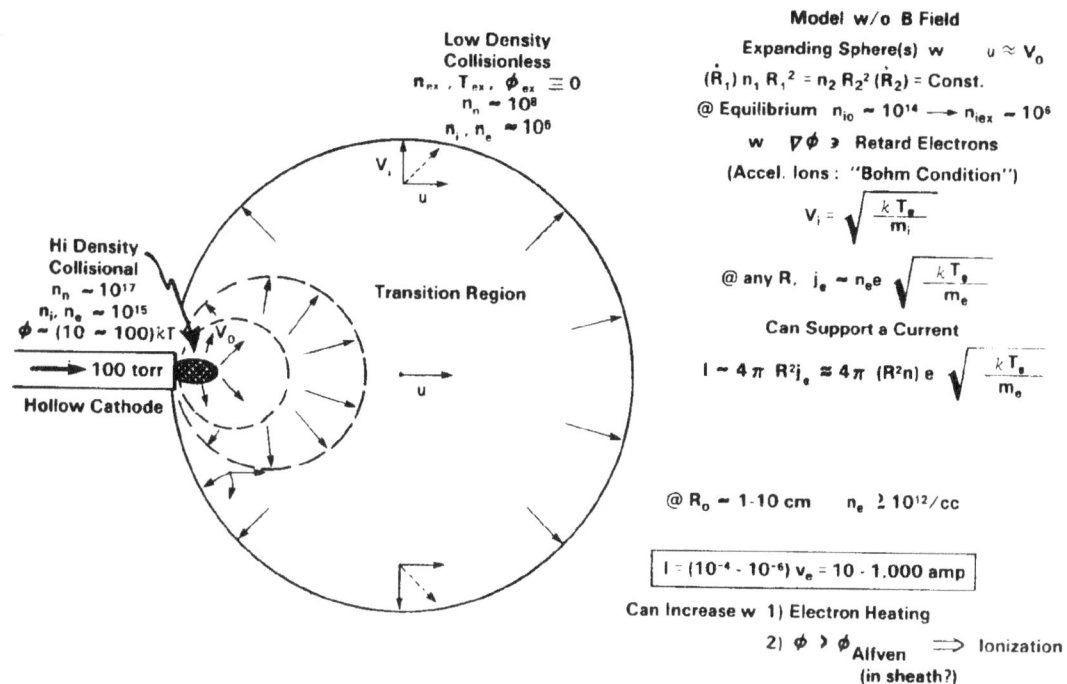

Figure 4.23 Electron Current Flow To/From the Ionosphere for PMG
Hollow Cathode Plasma Contactor

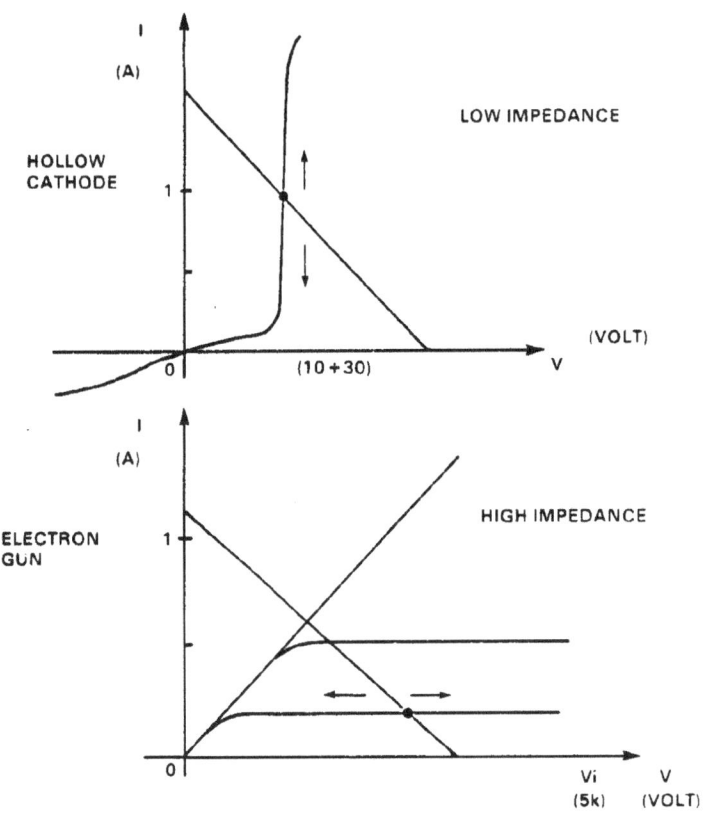

Figure 4.24 Comparison of the IV Characteristics of a Hollow Cathode and Electron Gun

	Electron Gun	**Hollow Cathode**
• **Current Range**	$I_e \leq 1A$	$I_e \geq 10$ A (Nominal)
• **Power Consumption**	~ 1 KW	~ 10 KW
• **Life Time**	Similar	Similar
• **Automatic Switching**	No	Yes
• **Main Applications**	Basic Science Exp. and Power	Low Impedance Coupling

Dissipation   **Thrusting and Power Generation**

Figure 4.25 Comparative Characteristics of an Electron Gun and a Hollow Cathode

The current passing through the tether can be controlled by any one of several methods, depending upon the type of plasma contactors used. For systems with passive conductors at both ends, control is by variable resistance, inserted between the tether and one of the plasma contactors. For systems using an electron gun as a plasma contactor, tether current is controlled by the current emitted by the electron gun. Unfortunately, these methods are very inefficient. They not only waste all of the I^2R power lost in the resistors, plasma sheaths (around the plasma contactors), and electron gun impedance, but they also transfer most of it as heat back into the spacecraft, where it may cause significant thermal control and heat rejection problems.

The basic equation of the current loop (circuit) is:

$$V_{IND} = IR + \Delta V_{LOW} + \Delta V_{UP} + \Delta V_{ION} + \Delta V_{LOAD} \; ;$$

where

$$
\begin{aligned}
V_{IND} &= \text{emf induced across the tether (volts),} \\
I &= \text{tether current (amps),} \\
R &= \text{resistance of the tether (ohms),} \\
\Delta V_{LOW} &= \text{voltage drop across the space charge region around the lower plasma} \\
& \quad \text{contactor (volts),} \\
\Delta V_{UP} &= \text{voltage drop across the space charge region around the upper plasma} \\
& \quad \text{contactor (volts),} \\
\Delta V_{ION} &= \text{voltage drop across the ionosphere (volts), and} \\
\Delta V_{LOAD} &= \text{voltage drop across a load (volts),}
\end{aligned}
$$

This equation simply states that the emf induced across the tether by its motion through the magnetic field is equal to the sum of all of the voltage drops in the circuit. The IR term in the equation is the voltage drop across the tether due to its resistance (according to Ohm's Law).

To provide an expression for the working voltage available to drive a load, this equation can be rewritten as:

$$\Delta V_{LOAD} = V_{IND} - IR - \Delta V_{LOW} - \Delta V_{UP} - \Delta V_{ION} \; .$$

The voltage drop across the space charge region (sheath, electron gun, or plasma cloud) at each tether end is caused by the impedance of that region. The voltage drop across the ionosphere is likewise due to its impedance. The problem with these equations is that the impedances of the charge regions around the tether ends are complex, nonlinear, and unknown functions of the tether current. The impedance of the ionosphere has not been clearly determined. Although some laboratory studies have been performed, and estimates made, detailed flight test measurements will have to be performed before these quantities can be clearly determined.

158

It has been calculated that the ionospheric impedance should be on the order of 1-20 ohms (Ref. 11). The highest impedance of the tether system are encountered at the space charge sheath regions around the upper and lower plasma contactors. Reducing these impedances will greatly increase the efficiency of the tether system in providing large currents. PMG data indicate that plasmas released from hollow cathode plasma contactors greatly reduce the sheath impedance between the contactors and the ambient plasma surrounding them. Although processes in these plasmas and in the ionosphere are not well understood and require continued study and evaluation through testing, preliminary indications are that feasible tether and plasma-contactor systems should be able to provide large induced currents.

As indicated earlier, the electric currents induced in such tether systems can be used to power loads on board the spacecraft equipped with them. They can also be used as primary power for the spacecraft. It has been calculated that electrodynamic tether systems should be capable of producing electrical power in the multikilowatt to possibly the megawatt range (Ref. 4, p. 161-184). Calculations a 200 KW system is given in figure 4.26.

There is a price to be paid for this electrical power, however. It is generated at the expense of spacecraft/tether orbital energy. This effect is described in detail in the next section .

In principle, electrodynamic tether systems can generate electrical power not only in Earth orbit, but also when they move through the magnetic fields of other planets and interplanetary space. The magnetic field in interplanetary space is provided by the solar wind, which is a magnetized plasma spiralling outward from the sun.

References 1 (p. 1-22 through 1-24, 3-49 through 3-65), 2, 4 (p. 153-184, 547-594), 10,11, and data from Dr. James McCoy (NASA/Johnson Space Center) are the primary references for this section.

TETHER LENGTH	20 KM (10 UP+10 DN)	WORKING TENSION	42 N
NOMINAL VOLTAGE	4 KV	WORKING ANGLE	17 DEG
RATED POWER	200 KW	RATED THRUST	25 N
PEAK POWER	500 KW	PEAK THRUST	>100 N

CONDUCTOR	#00 AWG ALUMINUM WIRE DIAMETER 9.3 MM @ 20°C RESISTANCE 8.4 OHNS @ 20°C 7.7 OHMS @ 0°C 7.1 OHMS @ -20°C	3640 KG
INSULATION	0.5 MM TEFLON (100 VOLTS/MIL)	
FAR END MASS	50 AMP HOLLOW CATHODE ASS'Y	278 KG
	(INCLUDING ELECTRONICS & CONTROL)	25 KG
TETHER CONTROLLER ELECTRONICS & MISC. HDWR.		94 KG
(POWER DISSIPATION LOSSES @ 1% = 2 KW)		
ARGON SUPPLY & CONTINGENCY RESERVE		163 KG
TOTAL		4,200 KG

TETHER DYNAMICS CONTROL	PASSIVE, IXB PHASING	
TETHER CURRENT/POWER CONTROL	DC IMPEDANCE MATCHING	
TETHER OUTSIDE DIAMETER	10.3 MM	
TETHER BALLISTIC DRAG AREA	206 SQ METERS	
DRAG FORCE @ 10^{-11} KG/M^3	12 N	.96 KW
(300 KM 1976 USSA-400 KM SOLAR MAX)		
I^2R LOSSES @ 200 KW		19.25 KW
HOLLOW CATHODE POWER		2.50 KW
IONOSPHERIC LOSS @ 50 AMP		1.25 KW
		—————
TOTAL PRIMARY LOSSES		23.96 KW

EFFICIENCY   ELECTRIC (177 KW NET @ 50 AMP/200KW) 88.5%
                OVERALL (201 MECH. TO 177 ELEC. KW) 88.1%

INCLUDING CONTROLLER/POWER PROCESSOR LOSSES @ 1%	2.00 KW
	—————
TOTAL (NET POWER OUT 175.0 KW)	25.96 KW

FINAL EFFICIENCY   ELECTRIC = 87.5%   OVERALL = 87.1%

Figure 4.26 Calculated Performance of an Electromagnetic Tether System

4.4.3 Thrusters

As mentioned in the previous two sections, electrodynamic tether systems can be used to generate thrust or drag. Consider the gravity-gradient-stabilized system in Earth orbit, for example. Its motion through the geomagnetic field induces an emf across the tether. When the current generated by this emf is allowed to flow through the tether, a force is exerted on the current (on the tether) by the geomagnetic field (see Figure 4.27). This force is given by:

$$\vec{F} = \int_{\substack{\text{along length of tether}}} (I\,\vec{dl}) \times \vec{B} = I \int_{\substack{\text{along length of tether}}} \vec{dl} \times \vec{B} \quad ;$$

where

$\vec{F}$ = force exerted on the tethe

I = tether current (amps),

$\vec{dl}$ = differential element of tet of positive current flow (

$\vec{B}$ = magnetic field strength (v

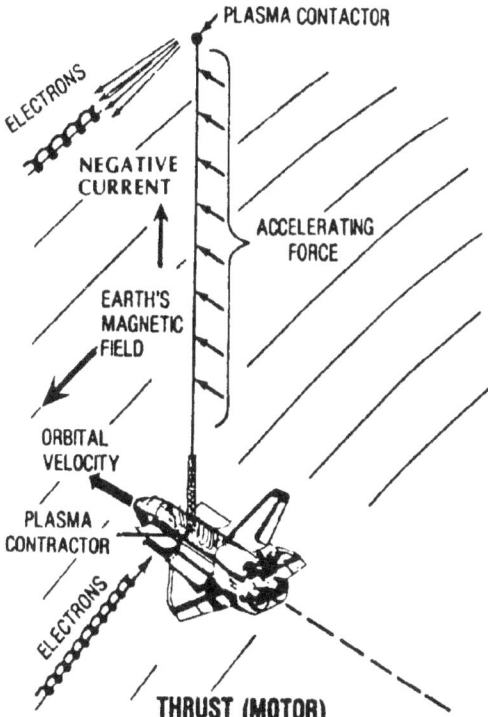

Figure 4.27 Thrust Generation With An Electrodynamic
Tether System

For the special case of a straight tether, this equation simplifies to:

$$\vec{F} = \vec{IL} \times \vec{B} \quad ;$$

where

$$\vec{L} = \text{tether length - a vector pointing in the direction of positive current flow (m).}$$

This equation for the electromagnetic force on a straight tether can also be written as:

$$F = ILB \sin \theta \quad ;$$

where

$$\theta = \text{angle between } \vec{L} \text{ and } \vec{B}.$$

Its maximum value occurs when the tether is perpendicular to the magnetic field.

Depending on the relative orientation of the magnetic field to the tether velocity, this force can have a component parallel to the velocity and one perpendicular to the velocity. Considering the parallel (inplane) component, whenever the current induced in the tether by the magnetic field is allowed to flow, this component of the force always acts to reduce the relative velocity between the tether system. In low Earth orbit, where the orbital velocity of the tether is greater than the rotational velocity of the geomagnetic field and they are rotating in the same direction, this force is a drag on the tether. This means that when electric power is generated by the system for on-board use, it is generated at the expense of orbital energy. If the system is to maintain its altitude, this loss must be compensated by rockets or other propulsive means.

When current from an on-board power supply is fed into the tether against the induced emf, the direction of this force is reversed. This force follows the same equation as before, but now the sign of the cross product is reversed, and the force becomes propulsive. In this way, the tether can be used as a thruster. Therefore, the same tether system can be used reversibly, as either an electric generator or as a thruster (motor). As always, however, there is a price to be paid. The propulsive force is generated at the expense of on-board electrical power.

It is necessary to distinguish between tether systems orbiting at subsynchronous altitudes, and those orbiting at altitudes greater than the synchronous altitude, where the sense of the relative velocity between the satellite and the magnetic field rest frame is reversed (often thought of in terms of a concept known as the "co-rotating field"). An analogous situation exists in orbits around Jupiter for altitudes greater than 2.2 Jovian radii from its center (the Jovian synchronous altitude: i.e., the altitude at which the rotational angular velocity of an orbiting satellite equals the rotational velocity of Jupiter and its magnetic field). Another analogous situation exists in interplanetary space if a spacecraft moves outward at a speed of 400 km/s). In such cases, dissipation of the induced electrical current would produce a thrust (not a

163

drag) on the tether. Again, the force acts to bring the relative velocity between the tether and the magnetic field rest frame to zero. In such cases, feeding current into the tether against the induced emf would produce a drag. When moving in a direction opposite to the direction of motion of the magnetic field, the effects would be reversed.

Systems have been proposed to operate reversibly as power and thrust generators (Ref. 4 and 10). Such systems could provide a number of capabilities. Calculations of the performance of a 200 KW system is given in figure 4.26.

In addition to the in-plane component, the electromagnetic force on the tether current generally also has an out-of-plane component (perpendicular to the tether velocity). For an orbiting tether system, the out-of-plane force component acts to change the orbital inclination, while doing no in-plane mechanical work on the tether and inducing no emf to oppose the flow of current in the tether. This makes electrodynamic tethers potentially ideal for orbital plane changes. Unlike rockets, they conserve energy during orbital plane changes. If the current is constant over a complete orbit, the net effect of this force is zero (since reversals in the force direction during the orbit cancel each other out). On the other hand, if a net orbital inclination change is desired, it can be produced by simply reversing the tether current at points in the orbit where the out-of-plane force reverses its direction, or by allowing a tether current to flow for only part of an orbit. Attention must be paid to this out-of-plane force when operating a tether alternately as a generator and thruster, and when operating a tether system which alternately generates and stores electrical energy. Strategies for using electrodynamic tethers to change orbits are shown in Section 5.0.

Electromagnetic forces also cause the tether to bow and produce torques on the tether system. These torques cause the system to tilt away from the vertical until the torques are balanced by gravity-gradient restoring torques. These torques produce in-plane and out-of-plane librations. The natural frequencies of in-plane and out-of-plane librations are $\sqrt{3}$ times the orbital frequency and twice the orbital frequency, respectively. Selective time phasing of the IL x B loading, or modulation of the tether current, will damp these librations. The out-of-plane librations are more difficult to damp because their frequency is twice the orbital frequency. Unless care is taken, day/night power generation/storage cycles (50/50 power cycles) can actively stimulate these librations. Careful timing of tether activities will be required to control all tether librations. Additional information on electromagnetic libration control issues is shown also in Section 5.0.

4.4.4 ULF/ELF/VLF Antennas

As discussed in Section 4.4.2, the movement of an Earth-orbiting electrodynamic tether system through the geomagnetic field gives rise to an induced current in the tether. One side effect of this current is that as the electrons are emitted from the tether back into the plasma, ULF, ELF, VLF electromagnetic waves are produced in the ionosphere (see Ref. 11).

In the current loop external to the tether, electrons spiral along the geomagnetic field lines and close at a lower layer of the ionosphere (see Figure 4.28). This current loop (or so-called "phantom loop") acts

164

as a large ULF, ELF, and VLF antenna. (The phantom loop is shown in Figure 4.29). The electromagnetic waves generated by this loop should propagate to the Earth's surface, as shown in Figure 4.30. The current flow generating these waves can be that induced by the geomagnetic field or can be provided by a transmitter on board the spacecraft so that the tether is in part an antenna.

Messages can be transmitted from the tether (antenna) by modulating the waves generated by the current loop. If the induced current is used to generate these waves, it is modulated by varying a series impedance or by turning an electron gun or hollow cathode on the lower tether end on and off at the desired frequency. If a transmitter is used, current is injected into the tether at the desired frequency.

The ULF, ELF, VLF waves produced in the ionosphere will be injected into the magnetosphere more efficiently than those from existing ground-based, man-made sources. It is believed that the ionospheric boundary may act as a waveguide, extending the area of effective signal reception far beyond the "hot spot" (area of highest intensity reception, with an estimated diameter of about 5000 km) shown in Figure 4.30. If this turns out to be the case, these waves may provide essentially instant worldwide communications, spreading over the Earth by ducting. Calculations have been performed, predicting that power levels of the order of 1 W by night and 0.1 W by day can be injected into the Earth-ionosphere transmission line by a 20-10 km tether with a current of the order of 10 A. Such tether systems would produce wave frequencies throughout the ULF (3-30 Hz) and ELF bands (30-300 Hz), and even into the VLF band (about 3000 Hz).

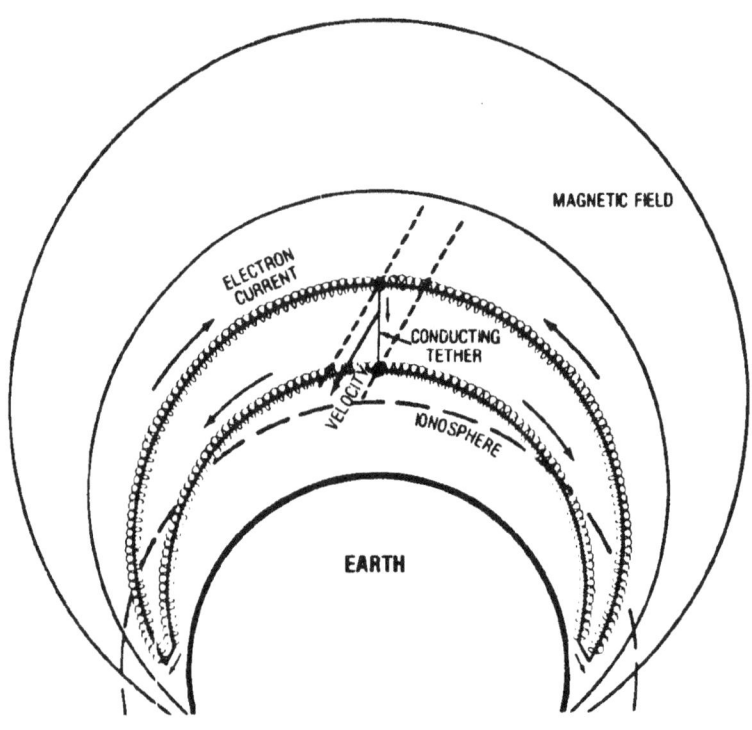

MAGNETIC FIELD

ELECTRON CURRENT

CONDUCTING TETHER

VELOCITY

IONOSPHERE

EARTH

Figure 4.28
Electron Paths in
the Electrodynamic
Tether Generator

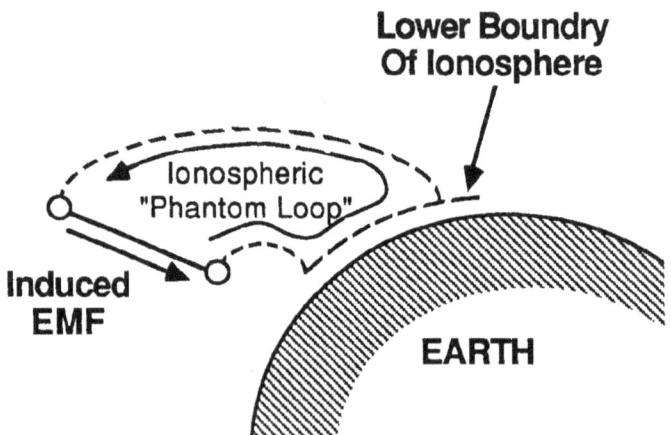

Lower Boundry
Of Ionosphere

Ionospheric
"Phantom Loop"

Induced
EMF

EARTH

Figure 4.29 The "Phantom Loop" of the ULF/ELF Tether Antenna

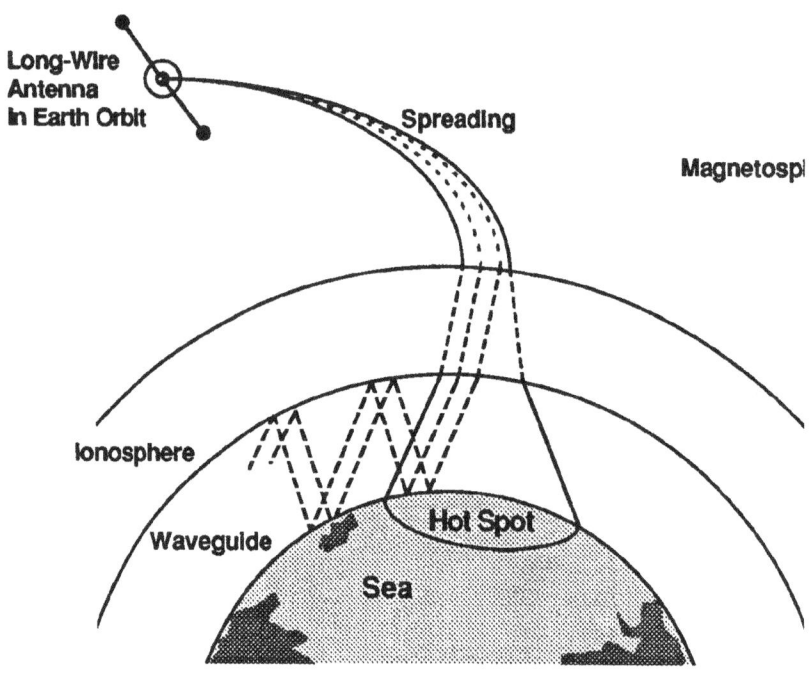

Figure 4.30 Propagation of ULF/ELF/VLF Waves To The Earth's
Surface From An Orbiting Tether Antenna

It should be noted that if the induced tether current is used to power the antenna, orbital energy will be correspondingly decreased. A means of restoring this orbital energy (such as rocket thrust) will be required for long missions.

4.4.5 Constellations

As mentioned earlier, electromagnetic forces exerted by the geomagnetic field on the current in orbiting tethers can be used in conjunction with gravity-gradient forces to stabilize two-dimensional constellations (see Figure 4.13). The force exerted on a current in a tether is exactly the force described in Section 4.4.3. The tether currents used in these constellations can be those induced by the geomagnetic field or those provided by on-board power supplies.

The basic concept is that gravity-gradient forces will provide vertical and overall attitude stability for the constellation, and electromagnetic forces will provide horizontal and shape stability (see Ref. 1, p.1-29, and 4, p. 150-203). This is accomplished in the quadrangular configuration by establishing the current direction in each of the vertical tethers such that the electromagnetic forces on them push the side arcs horizontally away from each other. Each side arc may be composed of a number of satellites connected in series by tethers. The current directions for the tethers on each side arc will be the same,

168

providing a consistent outward force. Large masses are placed at the top and bottom juncture points where the two sides join together. This provides additional stability for the constellation.

4.5   REFERENCES

1.   <u>Applications of Tethers in Space</u>, Workshop, Williamsburg, Virginia, 15-17 June 1983, Workshop Proceedings, NASA CP-2364 (Vol. 1), NASA CP-2365 (Vol. 2), March 1985.

2.   Beletskii, V. V. and Levin, E. M., "Dynamics of Space Tether Systems," <u>Advances in the Astronautical Sciences</u>, Vol. 83.

3.   Arnold, D. A., "The Behavior of Long Tethers in Space," <u>Journal of the Astronuatical Sciences</u>, Vol. 35, No. 1, p. 3-18, January-March, 1987.

4.   <u>Applications of Tethers in Space</u>, Workshop, Venice, Italy, 15-17 October 1985, Workshop Proceedings, NASA CP-2422 (Executive Summary, Vol. 1, Vol. 2), 1986.

5.   Carroll, J. A., <u>Guidebook for Analysis of Tether Applications</u>, Contract RH4-394049, Martin Marietta Corporation, Feb. 1985.

6.   Pearson, J., "Anchored Lunar Satellites for Cislunar Transportation and Communication," <u>Journal of the Astronautical Sciences</u>, Vol. 27, No. 1, p. 39-62, Jan.-Mar. 1979.

7.   Lorenzini, E. C., "Novel Tether-Connected Two-Dimensional Structures for Low Earth Orbits," <u>Journal of the Astronautical Sciences</u>, Vol. 36, No. 4, p. 389-405, Oct.-Dec. 1988.

8.   Greenwood, D. T., <u>Principles of Dynamics</u>, Prentice Hall, Inc., Englewood Cliffs, New Jersey, 1965.

9.   Tiesenhausen, G. von, ed., "The Roles of Tethers on Space Station," NASA-TM-86519, NASA/MSFC, Oct. 1985.

10.   McCoy, J. E., "Plasma Motor/Generator Reference System Designs for Power and Propulsion," AAS 86-229, Int. Conf. 1986.

11.   Grossi, M. D., "Spaceborne Long Vertical Wire as a Self-Powered ULF/ELF Radiator," <u>IEEE Journal of Oceanic Engineering</u>, Vol. OE-9, No. 3, p. 211-213, July 1984.

SECTION 5. TETHER DATA

5.1 General

This handbook would not be complete without providing the user with specific data and other information relevant to the analysis of tether applications. To the authors' knowledge, the best summarization of this data is contained in J. A. Carroll's <u>Guidebook for Analysis of Tether Applications,</u> published in 1985 under contract to the Martin Marietta Corporation. It provides a concise review of those technical areas which are essential to tether analyses. For the uninitiated, it is the first exposure they should have to ensure that they understand the broad implications of any application they might consider. From here, they can explore the many references given in the Bibliography.

The Guidebook is reproduced here in full, except for its bibliography which would be redundant. J. A. Carroll's introductory remarks and credits are presented below:

> This Guidebook is intended as a tool to facilitate initial analyses of proposed tether applications in space. The guiding philosophy is that at the beginning of a study effort, a brief analysis of all the common problem areas is far more useful than a detailed study in any one area. Such analyses can minimize the waste of resources on elegant but fatally flawed concepts, and can identify the areas where more effort is needed on concepts which do survive the initial analyses.

> In areas in which hard decisions have had to be made, the Guidebook is:

> Broad, rather than deep
> Simple, rather than precise
> Brief, rather than comprehensive
> Illustrative, rather than definitive

> Hence the simplified formulas, approximations, and analytical tools included in the Guidebook should be used only for preliminary analyses. For detailed analyses, the references with each topic and in the bibliography may be useful. Note that topics which are important in general but not particularly relevant to tethered system analysis (e.g., radiation dosages) are not covered.

> This Guidebook was presented by the author under subcontract RH4394049 with the Martin Marietta Corporation, as part of their contract NAS8-35499 (Phase II Study of Selected Tether Applications in Space) with the NASA Marshall Space Flight Center. Some of the material was adapted from references listed with the various topics, and this assisted the preparation greatly. Much of the other material evolved or was clarified in discussions with one or more of the following: Dave Arnold, James Arnold, Ivan Bekey, Guiseppe Colombo, Milt Contella, Dave Criswell, Don Crouch, Andrew Cutler, Mark Henley, Don Kessler, Harris Mayer, Jim McCoy, Bill Nobles, Tom O'Neil, Paul Penzo, Jack Slowey, Georg von Tiesenhausen, and Bill Thompson. The author is of course responsible for all errors, and would appreciate being notified of any that are found.

5.2 Generic Issues

MAJOR CONSTRAINTS IN MOMENTUM-TRANSFER APPLICATIONS

CONSTRAINT: / APPLICATION;	ORBIT BASICS	TETHER DYNAMICS	TETHER PROPERTIES	TETHER OPERATIONS
All types	Apside location	Forces on end masses	μmeteoroid sensitivity	Tether recoil at release
Librating		Tether can go slack		Facility attitude & "g"s variable
Spinning		High loads on payload		Retrieval can be difficult
Winching		High loads on payload		Extremely high power needed
Rendezvous	Orbit planes must match			Short launch & capture windows
Multi-stage	Dif. nodal regression			Waiting time between stages
High deltaV	Gravity losses	Control of dynamics	Tether mass & lifetime	Retrieval energy; Facility a alt.

MAJOR CONSTRAINTS WITH PERMANENTLY-DEPLOYED TETHERS

CONSTRAINTS: / APPLICATIONS:	ORBIT BASICS	TETHER DYNAMICS	TETHER PROPERTIES	TETHER OPERATIONS
All types	Aero. drag	Libration	Degradation, μmeteoroids & debris impact	Recoil & orbit changes after tether break
Electrodynamic	Misc changes in orbit	Plasma disturbances	High-voltage insulation	
Aerodynamic	Tether drag & heating			
Beanstalk (Earth)			Tether mass; debris impact	Consequences of failure
Gravity Use: Hanging / Spinning		Libration-Sensitive		<0.1 gee only, Docking awkward

5.3 Orbit Equations and Data

5.3.1 Orbits and Orbital Perturbations

KEY POINTS   Basic orbit nomenclature & equations are needed frequently in following pages. Comparison of tether & rocket operations requires orbit transfer equations.

The figures and equations at right are a summary of the aspects of orbital mechanics most relevant to tether applications analysis. For more complete and detailed treatments and many of the derivations, consult refs. 1-3.

The first equation in the box is known as the Vis Viva formulation, and to the right of it is the equation for the mean orbital angular rate, n. Much of the analysis of orbit transfer ΔVs and tether behavior follows from those two simple equations. Some analyses require a close attention to specific angular momentum, h, so an expression for h (for compact objects) is also given here.

In general, six parameters are needed to completely specify an orbit. Various parameter sets can be used (e.g., 3 position coordinates & 3 velocity vectors). The six parameters listed at right are commonly used in orbital mechanics. Note that when i=0, Ω becomes indeterminate (and unnecessary); similarly with ω when e=0. Also, i & Ω are here referenced to the central body's equator, as is usually done for Low Earth Orbit (LEO). For high orbits, the ecliptic or other planes are often used. This simplifies calculation of 3rd body effects.

NOTES   The effects of small ΔVs on near-circular orbits are shown at right. The relative effects are shown to scale: a ΔV along the velocity vector has a maximum periodic effect 4 times larger than that of the same, ΔV perpendicular to it (plus a secular effect in θ which the others don't have). Effects of oblique or consecutive ΔVs are simply the sum of the component effects. Note that out-of-plane ΔVs at a point other than a node also affect Ω.

For large ΔVs, the calculations are more involved. The perigee and apogee velocities of the transfer orbit are first calculated from the Vis Viva formulation and the constancy of h. Then the optimum distribution of plane change between the two ΔVs can be computed iteratively, and the required total ΔV found. Typically about 90% of the plane change is done at GEO.

To find how much a given in-plane tether boost reduces the required rocket ΔV, the full calculation should be done for both the unassisted and the tether-assisted rocket. This is necessary because the tether affects not only the perigee velocity, but also the gravity losses and the LEO/GEO plane change split. Each m/s of tether boost typically reduces the required rocket boost by 0.89 m/s (for hanging release) to 0.93 m/s (for widely librating release).

Note that for large plane changes, and large radius-ratio changes even without plane changes, 3-impulse "bi-elliptic" maneuvers may have the lowest total ΔV. Such maneuvers involve a boost to near-escape, a small plane and/or perigee-adjusting ΔV at apogee, and an apogee adjustment (by rocket or aerobrake) at the next perigee. In particular, this may be the best way to return aerobraking OTVs from GEO to LEO, if adequate time is available.

REFERENCES   1. A. E. Roy, Orbital Motion, Adam Hilger Ltd., Bristol, 1978.
2. Bate, Mueller, & White, Fundamentals of Astrodynamics, Dover Pub., 1971.
3. M. H. Kaplan, Modern Spacecraft Dynamics & Control, John Wiley & Sons, 1976.

Orbit & Orbit Transfer Equations

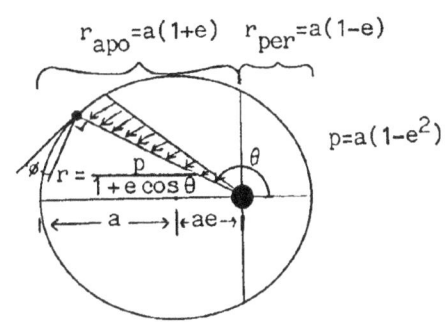

$r_{apo} = a(1+e)$ $r_{per} = a(1-e)$

$p = a(1-e^2)$

$r = \dfrac{p}{1 + e\cos\theta}$

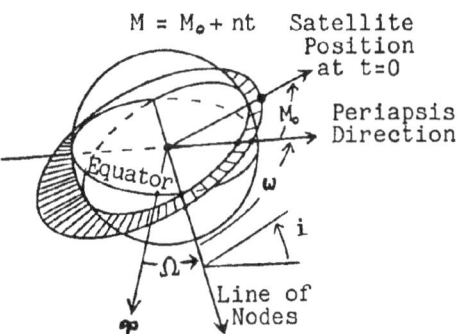

$M = M_o + nt$ Satellite Position at t=0

M_o

Periapsis Direction

Equator

ω

i

Ω

Line of Nodes

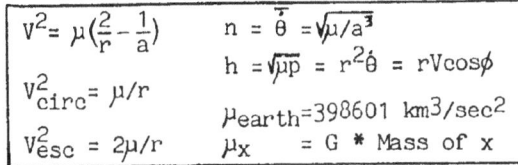

$v^2 = \mu\left(\dfrac{2}{r} - \dfrac{1}{a}\right)$	$n = \dot{\theta} = \sqrt{\mu/a^3}$
$v^2_{circ} = \mu/r$	$h = \sqrt{\mu p} = r^2\dot{\theta} = rV\cos\phi$
$v^2_{esc} = 2\mu/r$	$\mu_{earth} = 398601 \text{ km}3/\text{sec}^2$
	$\mu_x = G * \text{Mass of x}$

BASIC ORBIT EQUATIONS

ORBITAL ELEMENTS	
a	= semi-major axis
e	= eccentricity
i	= inclination
Ω	= long. of asc. node
ω	= argument of periapsis
M_o	= position at epoch

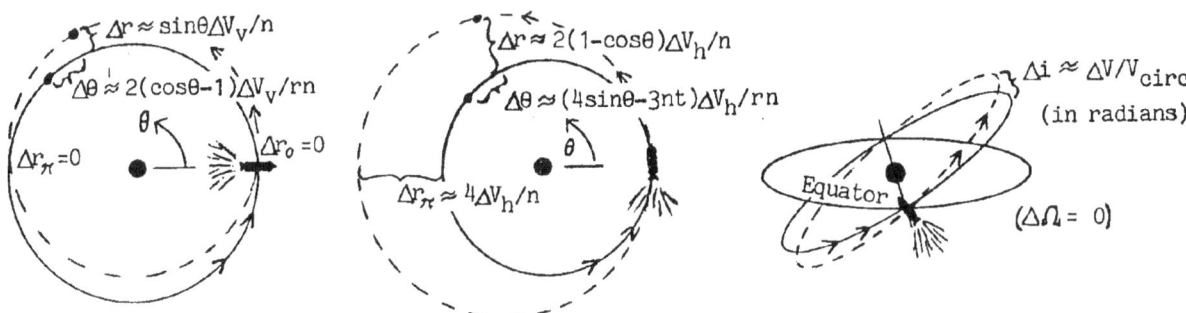

$\Delta r \approx \sin\theta \, \Delta V_V / n$

$\Delta\theta \approx 2(\cos\theta - 1)\Delta V_V / rn$

$\Delta r_\pi = 0$

$\Delta r_0 = 0$

$\Delta r \approx 2(1-\cos\theta)\Delta V_h / n$

$\Delta\theta \approx (4\sin\theta - 3nt)\Delta V_h / rn$

$\Delta r_\pi \approx 4\Delta V_h / n$

$\Delta i \approx \Delta V / V_{circ}$ (in radians)

Equator

$(\Delta\Omega = 0)$

Effects of Small ΔVs on Near-Circular Orbits

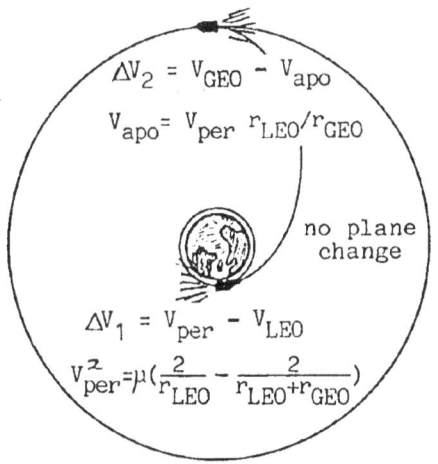

$\Delta V_2 = V_{GEO} - V_{apo}$

$V_{apo} = V_{per} \, r_{LEO}/r_{GEO}$

no plane change

$\Delta V_1 = V_{per} - V_{LEO}$

$V^2_{per} = \mu\left(\dfrac{2}{r_{LEO}} - \dfrac{2}{r_{LEO} + r_{GEO}}\right)$

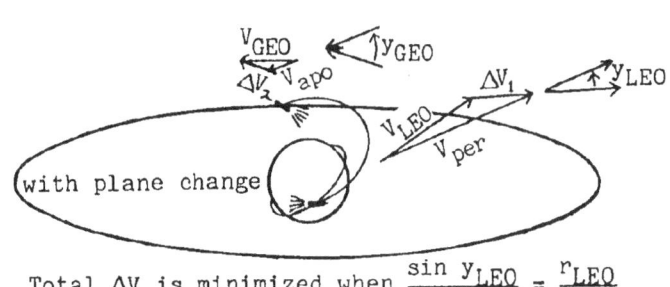

V_{GEO} y_{GEO}

V_{apo} ΔV_2

ΔV_1 y_{LEO}

V_{LEO}

V_{per}

with plane change

Total ΔV is minimized when $\dfrac{\sin y_{LEO}}{\sin y_{GEO}} = \dfrac{r_{LEO}}{r_{GEO}}$

Large Orbit Transfers (e.g., LEO—GEO)

5.3.2 Orbital Perturbations

KEY POINTS   Differential nodal regression severely limits coplanar rendezvous windows. Apsidal recession affects STS deboost requirements from elliptical orbits. Third bodies can change the orbit plane of high-orbit facilities.

The geoid (earth's shape) is roughly that of a hydrostatic-equilibrium oblate ellipsoid, with a 296:297 polar:equatorial radius ratio. There are departures from this shape, but they are much smaller than the 1:297 oblateness effect and have noticeable effects only on geosynchronous and other resonant orbits.

The focus here is on oblateness, because it is quite large and because it has large secular effects on Ω and ω for nearly all orbits. (Oblateness also affects n, but this can usually be ignored in preliminary analyses.) As shown at right, satellites orbiting an oblate body are attracted not only to its center but also towards its equator. This force component imposes a torque on all orbits that cross the equator at an angle, and causes the direction of the orbital angular momentum vector to regress as shown.

$\dot{\Omega}$ is largest when i is small, but the plane change associated with a given $\Delta\Omega$ varies with sin(i). Hence the actual plane change rate varies with sin(i)cos(i), or sin(2i), and is highest near 45°. For near-coplanar rendezvous in LEO, the required out-of-plane ΔV changes by 78sin(2i) m/s for each phasing "lap". This is independent of the altitude difference (to first order), since phasing & differential nodal regression rates both scale with Δa. Hence even at best a rendezvous may require an out-of-plane ΔV of 39 m/s. At other times, out-of-plane ΔVs of $2sin(i)sin(\Delta\Omega/2)V_{circ}$ (= up to 2 V_{circ}!) are needed.

NOTES   The linkage between phasing and nodal regression rates is beneficial in some cases: if an object is boosted slightly and then allowed to decay until it passes below the boosting object, the total $\Delta\Omega$ is nearly identical for both. Hence recapture need not involve any significant plane change.

Apsidal recession generally has a much less dominant effect on operations, since apsidal adjustments (particularly of low-e orbits) involve much lower ΔVs than nodal adjustments. However, tether payload boosts may often be done from elliptical STS orbits, and perigee drift may be an issue. For example, OMS deboost requirements from an elliptical STS orbit are tonnes lower (and payload capability much higher) if perigee is near the landing site latitude at the end of the mission. Perigee motion relative to day/night variations is also important for detailed drag calculations, and for electrodynamic day-night energy storage (where it smears out and limits the eccentricity-pumping effect of a sustained day-night motor-generator cycle).

Just as torques occur when the central body is non-spherical, there are also torques when the satellite is non-spherical. These affect the satellite's spin axis and cause it to precess around the orbital plane at a rate that depends on the satellite's mass distribution and spin rate.

In high orbits, central-body perturbations become less important and 3rd-body effects more important. In GEO, the main perturbations (~47 m/s/yr) are caused by the moon and sun. The figure at right shows how to estimate these effects, using the 3rd body orbital plane as the reference plane.

REFERENCES 1. A. E. Roy, Orbital Motion, Adam Hilger Ltd., Bristol, 1978.
2. Bate, Mueller, & White, Fundamentals of Astrodynamics, Dover Pub., 1971.

Orbital Perturbations

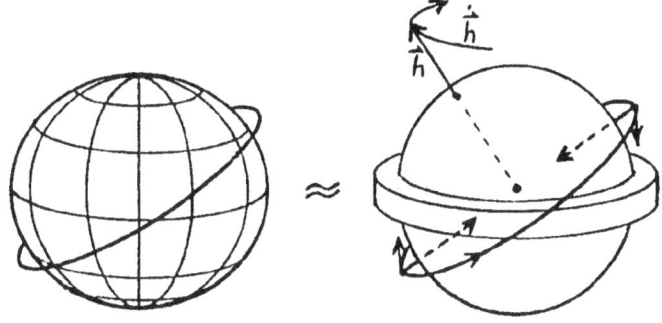

OBLATENESS CAUSES LARGE
SECULAR CHANGES IN Ω & ω:

$\dot{\Omega}$: up to 1 rad./week in LEO

$\dot{\omega}$: up to 2 rad./week in LEO

Nodal Regression in LEO:

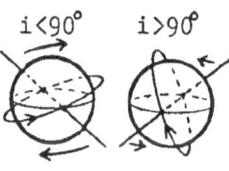

$$\dot{\Omega} \approx \frac{-63.6 \cos i \text{ rad/yr}}{(a/r_e)^{3.5} (1-e^2)^2}$$

$$(r_e = 6378 \text{ km})$$

For sun-synchronous orbits: $(i=100° \pm 4°)$

$$\cos i \approx -.0988(a/r_e)^{3.5}(1-e^2)^2$$

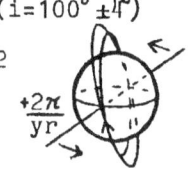

For coplanar low-ΔV rendezvous
between 2 objects ($e_1 = e_2 \approx 0$, $i_1 = i_2$),
nodal coincidence intervals are:

$$\Delta t_{nc} \approx \frac{180 \ (\bar{a}/re)^{4.5}}{\Delta a \ |\cos i|} \text{ km·yrs}$$

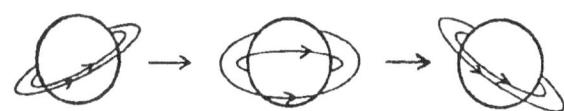

Apsidal recession in LEO:

$$\dot{\omega} \approx \frac{63.6(2 - 2.5 \sin^2 i)}{(a/r_e)^{3.5} (1-e^2)^2} \text{ rad/yr}$$

$i<63.4°$ $i=63.4°$ $i>63.4°$

Motion of the longitude of
perigee with respect to the
sun's direction ("noon") is:

$$\bar{\omega}_s = \dot{\omega} + \dot{\Omega} - 2\pi/yr$$

$$\dot{\Omega}_{s3} = -.75 \cos i_{s}, \ \mu_3 \ /n_s \ r_3^3$$

"Smeared out" 3rd body

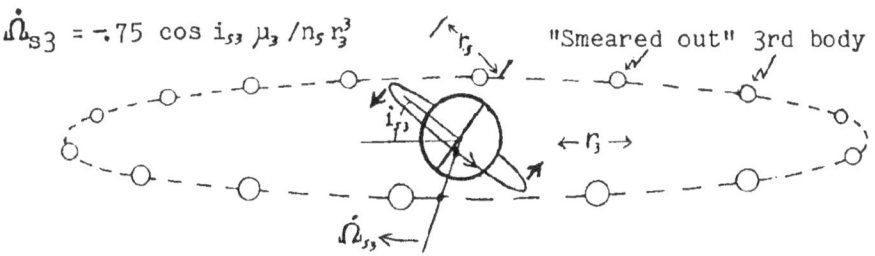

Third-Body Perturbations (non-resonant orbits)

177

5.3.3 Aerodynamic Drag

KEY POINTS   Tether drag affects tether shape & orbital life; atomic oxygen degrades tethers. Out-of-plane drag component can induce out-of-plane tether libration. The main value of payload boosting by tether is the increased orbital life. Unboosted orbital life of space facilities is affected by tether operations.

The figure at right shows the orbiter trolling a satellite in the atmosphere, as is planned for the 2nd TSS mission in the late 1980s. The tether drag greatly exceeds that on the end-masses and should be estimated accurately. The drag includes a small out-of-plane component that can cause ϕ-libration.

Tether drag is experienced over a range of altitudes, over which most of the terms in the drag equation vary: the air density ρ, the airspeed V_{rel}, and the tether width & angle of attack. In free-molecular flow, C_L is small, and C_D (if based on $A_\perp$) is nearly constant at 2.2. (CD rises near grazing incidence, but then $A_\perp$ is low.)

Only ρ varies rapidly, but it varies in a way which lends itself to simple approximations. Empirical formulae have been developed by the author and are shown at right. They give values that are usually within 25% of ref. 1, which is still regarded as representative for air density as a function of altitude & exosphere temperature. These estimates hold only for $\rho > 1E-14$, beyond which helium & hydrogen dominate & the density scale height H increases rapidly.

NOTES   Note that over much of LEO, atomic oxygen is the dominant species. Hyperthermal impact of atomic oxygen on exposed surfaces can cause rapid degradation, and is a problem in low-altitude applications of organic-polymer tethers.

The space age began in 1957 at a 200-yr high in sunspot count. A new estimate of mean solar cycle temperatures (at right, from ref. 2), is much lower than earlier estimates. Mission planning requires both high & mean estimates for proper analysis. Ref. 2 & papers in the same volume discuss models now in use.

If the tether length L is <<H, the total tethered system drag can be estimated from the total $A_\perp$ & the midpoint V & ρ. If L>>H, the top end can be neglected, the bottom calculated normally, and the tether drag estimated from $1.1\rho_{bottom}$ * tether diameter * H * V_{rel}^2, with H & V_{rel} evaluated one H above the bottom of the tether. For L between these cases, the drag is bounded by these cases.

As shown at right, the orbital life of more compact objects (such as might be boosted or deboosted by tether) can be estimated analytically if T_{ex} is known. For circular orbits with the same r, V_{rel} & $\bar{\rho}$ both vary with i, but these variations tend to compensate & can both be ignored in first-cut calculations.

The conversion of elliptical to "equal-life" circular orbits is an empirical fit to an unpublished parametric study done by the author. It applies when apsidal motions relative to the equator and relative to the diurnal bulge are large over the orbital life; this usually holds in both low & high-i orbits. For a detailed study of atmospheric drag effects, ref. 3 is still useful.

REFERENCES   1. U.S. Standard Atmosphere Supplements, 1966. ESSA/NASA/USAF, 1966.
2. K. S. W Champion, "Properties of the Mesosphere and Thermosphere and Comparison with CIRA 72", in The Terrestrial Upper Atmosphere, Champion and Roemer, ed.; Vol 3, #1 of Advances in Space Research, Pergamon, 1983.

178

3. D. G. King-Hele, Theory of Satellite Orbits in an Atmosphere, Butterworths, London, 1964.

Aerodynamic Drag

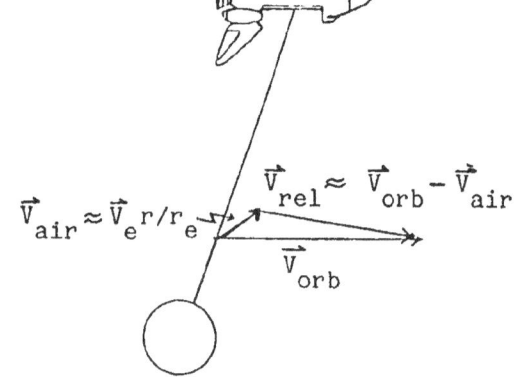

$$\vec{V}_{air} \approx \vec{V}_e r/r_e$$

$$\vec{V}_{rel} \approx \vec{V}_{orb} - \vec{V}_{air}$$

$$\vec{V}_{orb}$$

$$F_{drag} = .5 \int\!\!\int \rho \, C_D \, V_{rel}^2 \, \text{Width} \, \delta r$$

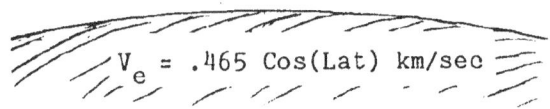

$$V_e = .465 \, \text{Cos(Lat) km/sec}$$

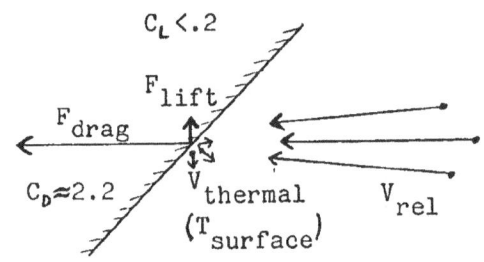

$C_L < .2$

F_{lift}

F_{drag}

$C_D \approx 2.2$

$V_{thermal}$ ($T_{surface}$)

V_{rel}

Lift & Drag in Free-Molecular Flow

$$(\bar{\lambda} \gg D_{tether}; \quad \bar{\lambda} \approx \frac{10^{-7} \, \text{kg/m}^2}{\rho})$$

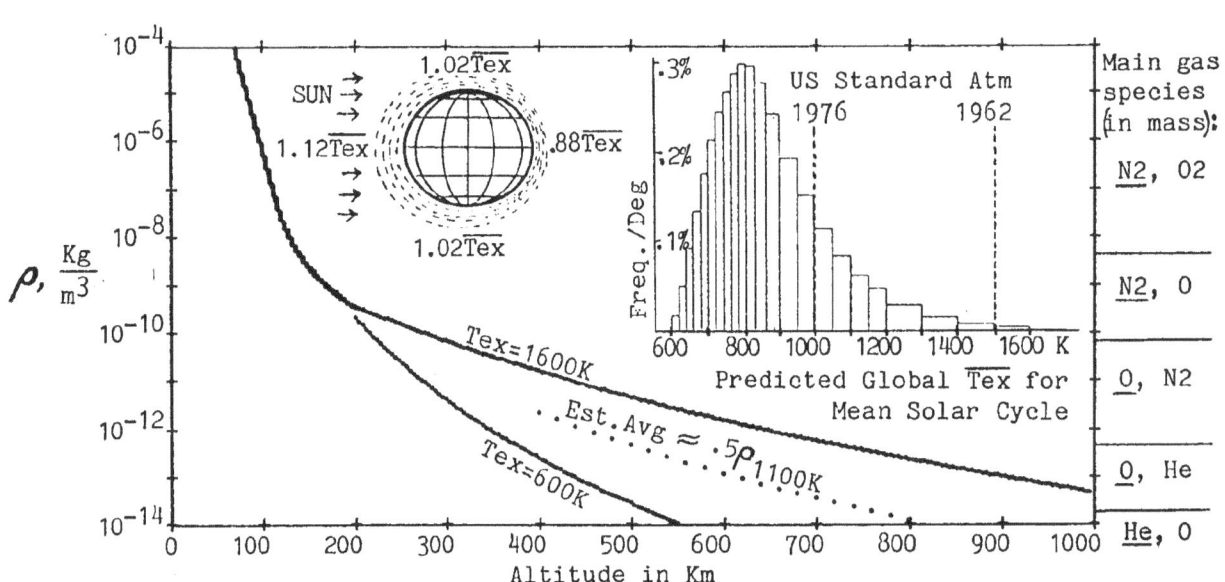

Air Density as Function of Altitude & Exosphere Temperature

$70 < \text{Alt} < 118: \ \rho \approx 11\exp(-\text{Alt}/6)$	$-\rho/\dot{\rho} = H = 6 \, (\text{km})$
$118 < \text{Alt} < 200: \ \rho \approx (\text{Alt}-95)^{-3}/2600$	$H = (\text{Alt}-95)/3$
$200 < \text{Alt}$ &: $\ \rho \approx \dfrac{1.47\text{E-}16 \ \text{Tex}(3000-\text{Tex})}{(1+2.9(\text{Alt}-200)/\text{Tex})^{10}}$	$H = .1(\text{Alt}-200) + \text{Tex}/29$
$\rho > 1\text{E-}14$	

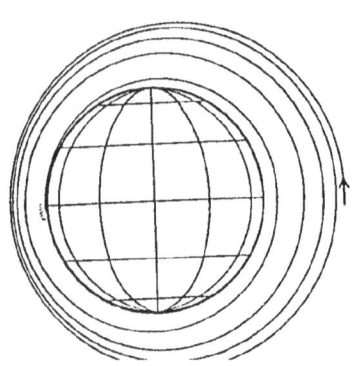

$$\text{Circular Orbit Life} \approx \frac{.15 \ \text{m}^2\text{yr}}{\text{kg}} \ \frac{M}{C_D A} \ \frac{(1 + 2.9(r-6578)/\text{Tex})^{11}}{3000 - \text{Tex}}$$

$(-14 < \text{Log}\rho < -10)$

$$\text{Equal-Life Circ. Alt.} \approx \text{Perigee} + \frac{\text{Apo} - \text{Per}}{2 + .154(\text{Apo-Per})/H_{\text{Per}}}$$

180

5.3.4 Thermal Balance

KEY POINTS   Aerothermal heating of tethers is severe at low altitudes (<120 km). Tether temperature affects strength, toughness, & electrical conductivity. Extreme thermal cycling may degrade pultruded composite tethers. "View factors" are also used in refined micrometeoroid risk calculations.

Preliminary heat transfer calculations in space are often far simpler than typical heat transfer calculations on the ground, since the complications introduced by convection are absent. However the absence of the "clamping" effect of large convective couplings to air or liquids allows very high or low temperatures to be reached, and makes thermal design important.

At altitudes below about 140 km in LEO, aerodynamic heating is the dominant heat input on surfaces facing the ram direction. The heating scales with ρ as long as the mean free path λ is much larger than the object's radius. It is about equal to the energy dissipated in stopping incident air molecules. In denser air, shock & boundary layers develop. They shield the surface from the incident flow and make $\dot{Q}$ rise slower as ρ increases further. (See ref 1.)

Because tethers are narrow, they can be in free molecular flow even at 100 km, and may experience more severe heating than the (larger) lower end masses do. Under intense heating high temperature gradients may occur across non-metallic tethers. These gradients may cause either overstress or stress relief on the hot side, depending on the sign of the axial thermal expansion coefficient.

NOTES   At higher altitudes the environment is much more benign, but bare metal (low-emittance) tethers can still reach high temperatures when resistively heated or in the sun, since they radiate heat poorly. Silica, alumina, or organic coatings >1 μm thick can increase emittance and hence reduce temperatures. The temperature of electrodynamic tethers is important since their resistance losses (which may be the major system losses) scale roughly with T_{abs}.

For a good discussion of solar, albedo, and long wave radiation, see ref. 2. The solid geometry which determines the gains from these sources is simple but subtle, and should be done carefully. Averaged around a tether, earth view-factors change only slowly with altitude & attitude, and are near 0.3 in LEO.

Surface property changes can be an issue in long-term applications, due to the effects of atomic oxygen, UV & high-energy radiation, vacuum, deposition of condensable volatiles from nearby surfaces, thermal cycling, etc. Hyperthermal atomic oxygen has received attention only recently, and is now being studied in film, fiber, and coating degradation experiments on the STS & LDEF.

Continued thermal cycling over a wide range (such as shown at bottom right) may degrade composite tethers by introducing a maze of micro-cracks. Also, temperature can affect the strength, stiffness, shape memory, and toughness of tether materials, and hence may affect tether operations and reliability.

REFERENCES   1. R. N. Cox & L.F. Crabtree, Elements of Hypersonic Aerodynarnics, The English Universities Press Ltd, London, 1965. See esp. Ch 9, "Low Density Effects"
2. F. S. Johnson, ed., Satellite Environment Handbook, Second Edition, Stanford University Press, 1965. See chapters on solar & earth thermal radiation.
3. H. C. Hottel, "Radiant Heat Transmission," Chapter 4 of W.H. McAdams, HEAT TRANSMISSION, 3rd edition, McGraw-Hill, New York, 1954, pp. 55-125.

Thermal Balance

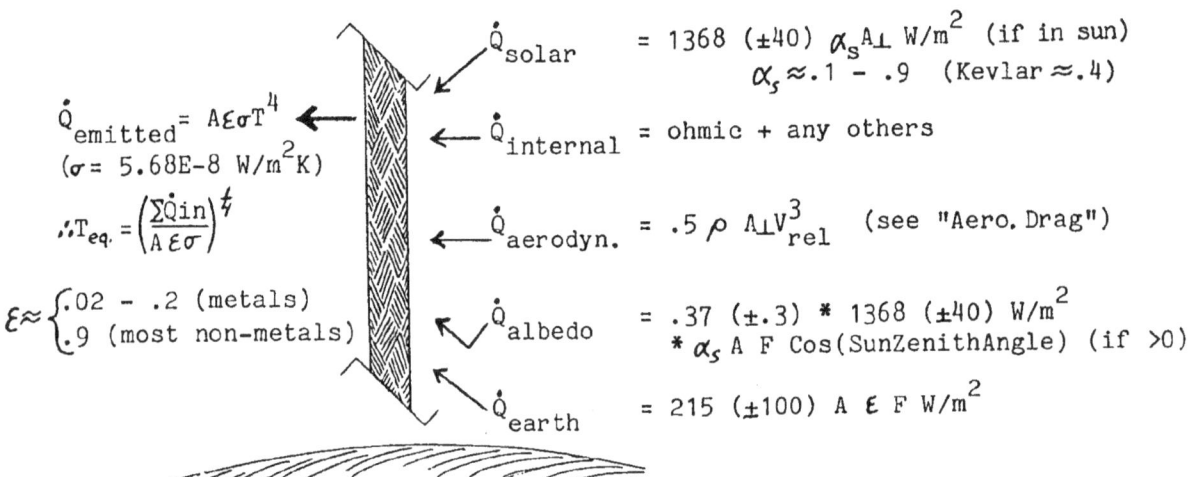

$$\dot{Q}_{solar} = 1368\ (\pm 40)\ \alpha_s A_\perp\ W/m^2\ \text{(if in sun)}$$
$$\alpha_s \approx .1 - .9\ \text{(Kevlar} \approx .4)$$

$$\dot{Q}_{emitted} = A\varepsilon\sigma T^4$$
$$(\sigma = 5.68E-8\ W/m^2K)$$

$$\dot{Q}_{internal} = \text{ohmic + any others}$$

$$\therefore T_{eq.} = \left(\frac{\Sigma\dot{Q}in}{A\varepsilon\sigma}\right)^{\frac{1}{4}}$$

$$\dot{Q}_{aerodyn.} = .5\ \rho\ A_\perp V_{rel}^3\ \text{(see "Aero. Drag")}$$

$$\varepsilon \approx \begin{cases} .02 - .2\ \text{(metals)} \\ .9\ \text{(most non-metals)} \end{cases}$$

$$\dot{Q}_{albedo} = .37\ (\pm .3)\ *\ 1368\ (\pm 40)\ W/m^2$$
$$*\ \alpha_s\ A\ F\ Cos(SunZenithAngle)\ \text{(if} >0)$$

$$\dot{Q}_{earth} = 215\ (\pm 100)\ A\ \varepsilon\ F\ W/m^2$$

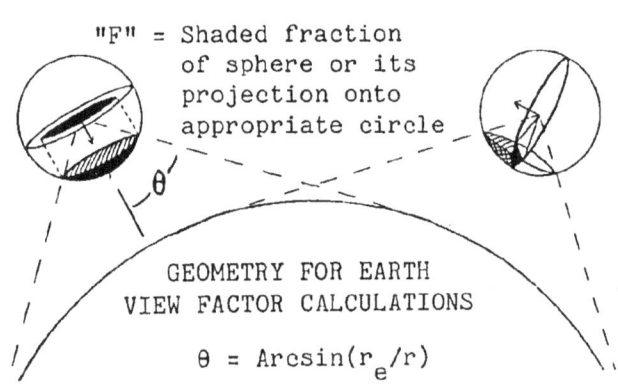

"F" = Shaded fraction of sphere or its projection onto appropriate circle

GEOMETRY FOR EARTH VIEW FACTOR CALCULATIONS

$$\theta = Arcsin(r_e/r)$$

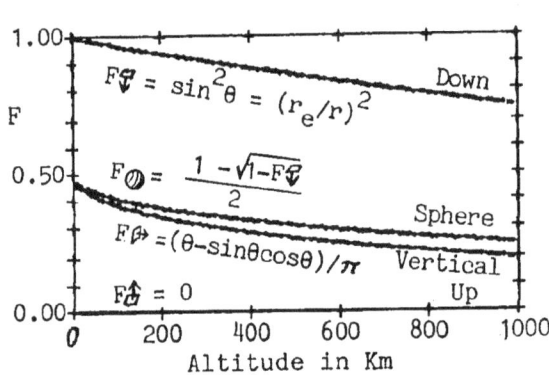

$$F\ \text{(Down)} = sin^2\theta = (r_e/r)^2$$

$$F\ \text{(Sphere)} = \frac{1 - \sqrt{1-F\downarrow}}{2}$$

$$F\ \text{(Vertical)} = (\theta - sin\theta cos\theta)/\pi$$

$$F\ \text{(Up)} = 0$$

Earth Viewfactors in LEO

$$F\ \text{||} = F\ \text{(Vertical)}$$
$$F\ \text{(mesh)} \approx (F\uparrow + F\downarrow + 2F\rightarrow)/4$$
$$F\ \text{||} < F\ \text{/} < F\ \text{(mesh)}$$

Earth Viewfactors for Tethers

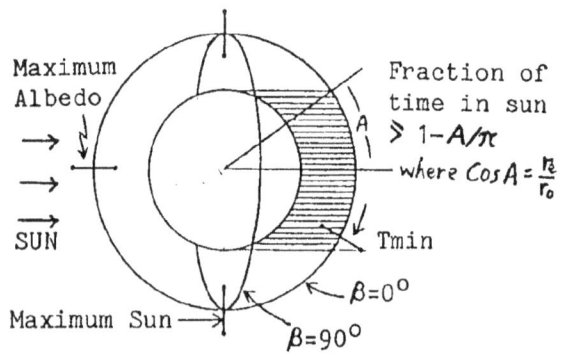

Maximum Albedo

SUN

Fraction of time in sun $\geq 1 - A/\pi$ where $Cos A = \frac{r_e}{r_o}$

Tmin

$\beta = 0°$
$\beta = 90°$

Maximum Sun

Inclination	β Range
0°	0-23.5°
28.5°	0-52°
>66.5°	0-90°

(β = Sun out-of-plane angle)

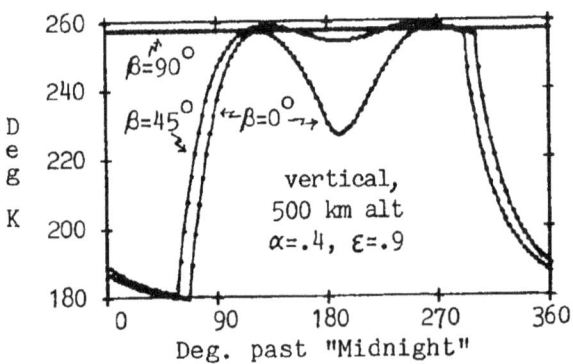

$\beta=90°$
$\beta=45°$
$\beta=0°$

vertical, 500 km alt $\alpha=.4$, $\varepsilon=.9$

Deg. past "Midnight"

Tether Temperature Over 1 Orbit

5.3.5 Micrometeoroids and Debris

KEY POINTS   Micrometeoroids can sever thin tethers & damage tether protection/insulation. Orbiting debris can sever tethers of any diameter.

At the start of the space age, estimates of meteoroid fluxes varied widely. Earth was thought to have a dust cloud around it, due to misinterpretation of data such as microphone noise caused by thermal cycling in spacecraft. By the late 1960s most meteoroids near earth were recognized to be in heliocentric rather than geocentric orbit. The time-averaged flux is mostly sporadic, but meteor showers can be dominant during their occurrence.

There is a small difference between LEO and deep-space fluxes, due to the focusing effect of the earth's gravity (which increases the velocity & flux), and the partial shielding provided by the earth & "sensible" atmosphere. For a typical meteoroid velocity of 20 km/sec, these effects combine to make the risk vary as shown at right in LEO, GEO, and beyond. The picture of a metal plate after hypervelocity impact is adapted from ref. 3.

The estimated frequency of sporadic meteoroids over the range of interest for most tether applications is shown by the straight line plot at right, which is adapted from ref. 4 & based on ref. 1. (Ref 1 is still recommended for design purposes.) For masses <1E-6 gm (<0.15 mm diam. at an assumed density of 0.5), the frequency is lower than an extension of that line, since several effects clear very small objects from heliocentric orbits in geologically short times.

NOTES   Over an increasing range of altitudes and particle sizes in LEO, the main impact hazard is due not to natural meteoroids but rather to man-made objects. The plots at right, adapted from refs 4 & 5, show the risks presented by the 5,000 or so objects tracked by NORAD radars (see ref. 6). A steep "tail" in the 1995 distribution is predicted since it is likely that several debris-generating impacts will have occurred in LEO before 1995. Such impacts are expected to involve a 4-40 cm object striking one of the few hundred largest objects and generating millions of small debris fragments.

Recent optical detection studies which have a size threshold of about 1 cm indicate a population of about 40,000 objects in LEO. This makes it likely that debris-generating collisions have already occurred. Studies of residue in small surface pits on the shuttle and other objects recovered from LEO indicate that they appear to be due to titanium, aluminum, and paint fragments (perhaps flaked off satellites by micrometeoroid hits). Recovery of the Long Duration Exposure Facility (LDEF) later this year should improve this database greatly, and will provide data for LEO exposure area-time products comparable to those in potential long-duration tether applications.

REFERENCES 1. Meteoroid Environment Model—1969 [Near Earth to Lunar Surface], NASA SP-8013, March 1969.
2. Meteoroid Environment Model—1970 [Interplanetary and Planetary], NASA SP-8038, October 1970.
3. Meteoroid Damage Assessment, NASA SP-8042, May 1970. Shows impact effects.
4. D. J. Kessler, "Sources of Orbital Debris and the Projected Environment for Future Spacecraft", in J. of Spacecraft & Rockets, Vol 18 #4, Jul-Aug 1981.
5. D. J. Kessler, Orbital Debris Environment for Space Station, JSC-20001, 1984.
6. CLASSY Satellite Catalog Compilations. Issued monthly by NORAD/J5YS, Peterson Air Force Base, CO 80914.

Micrometeoroids & Debris

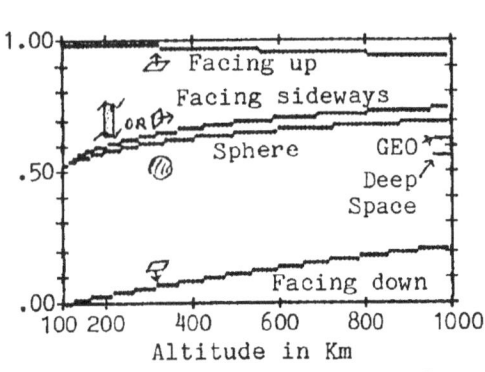

Relative μm Risks in LEO

$$[RelRisk \approx (1 - F_{earth})(.57 + .43r_e/r)]$$

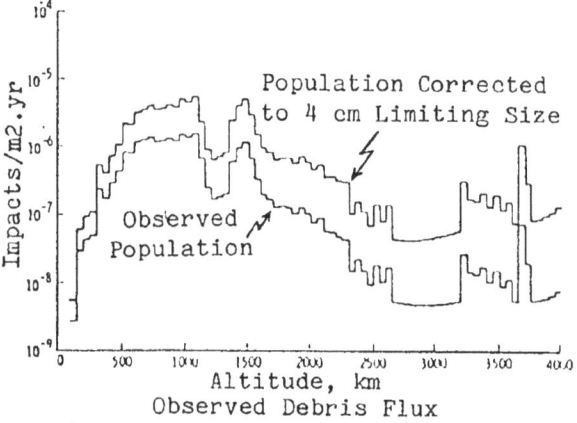

Observed Debris Flux
(corrected to 4-cm limiting size)

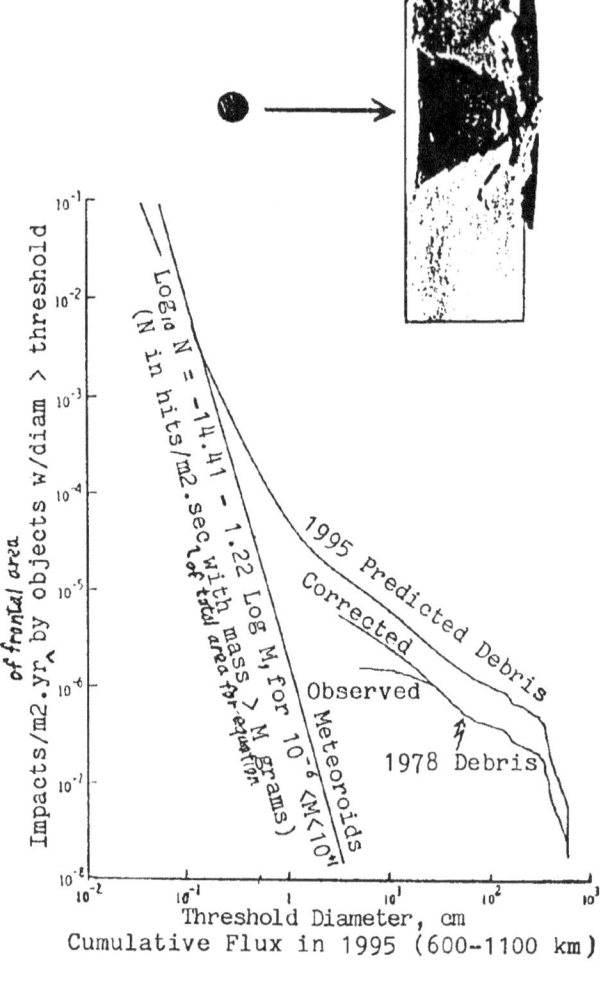

Cumulative Flux in 1995 (600-1100 km)

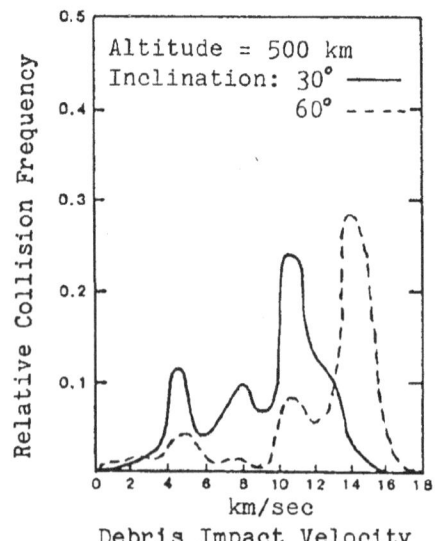

Debris Impact Velocity

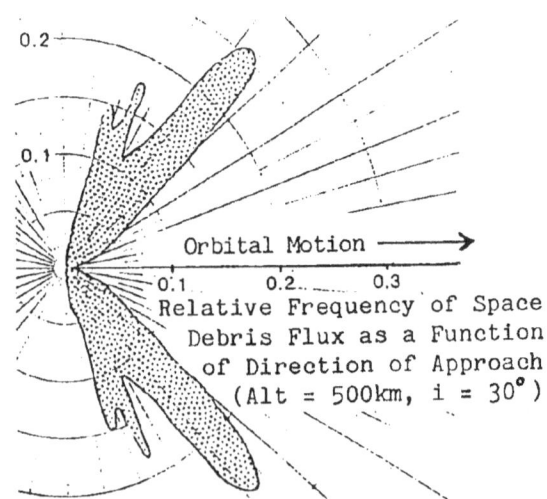

Relative Frequency of Space
Debris Flux as a Function
of Direction of Approach
(Alt = 500km, i = 30°)

184

5.4 Tether Dynamics and Control

5.4.1 Gravity Gradient Effects

KEY POINTS "Microgee" environments are possible only in small regions (~5 m) of a LEO facility. Milligee-level gravity is easy to get & adequate for propellant settling, etc.

The figure at right shows the reason for gravity-gradient effects. The long tank-like object is kept aligned with the local vertical, so that the same end always faces the earth as it orbits around it. If one climbs from the bottom to the top, the force of gravity gradually decreases and the centrifugal force due to orbital motion increases. Those forces cancel out only at one altitude, which is (nearly but not exactly) the altitude of the vehicle's center of mass.

At other locations an object will experience a net force vertically away from the center of mass (or a net acceleration, if the object is allowed to fall). This net force is referred to as the "gravity-gradient force." (But note that 1/3 of the net force is actually due to a centrifugal force gradient!) Exact and approximate formulas for finding the force on an object are given at right.

The force occurs whether or not a tether is present, and whether or not it is desirable. Very-low-acceleration environments, which are needed for some types of materials processing and perhaps for assembling massive structures, are only available over a very limited vertical extent, as shown at right. Putting a vehicle into a slow retrograde spin can increase the "height" of this low-gee region, but that then limits the low-gee region's other in-plane dimension.

NOTES Since gravity gradients in low orbits around various bodies vary with μ/r^3, the gradients are independent of the size of the body, and linearly dependent on its density. Hence the gradients are highest (.3-.4 milligee/km) around the inner planets and Earth's moon, and 60-80% lower around the outer planets. In higher orbits, the effect decreases rapidly (to 1.6 microgee/km in GEO).

The relative importance of surface tension and gravity determines how liquids behave in a tank, and is quantified with the Bond number, $B_o = \rho a r/\sigma$. If $B_o > 10$, liquids will settle, but higher values ($B_o = 50$) are proposed as a conservative design criterion. On the other hand, combining a small gravity gradient effect ($B_o < 10$) with minimal surface-tension fluid-management hardware may be more practical than either option by itself. Locating a propellant depot at the end of a power tower structure might provide an adequate gravity-gradient contribution. If higher gravity is desired, but without deploying the depot, another option is to deploy an "anchor" mass on a tether, as shown at right.

Many nominally "zero-gee" operations such as electrophoresis may actually be compatible with useful levels of gravity (i.e., useful for propellant settling, simplifying hygiene activities, keeping objects in place at work stations, etc.). This needs to be studied in detail to see what activities are truly compatible.

REFERENCES 1. D. Arnold, "General Equations of Motion," Appendix A of Investigation of Electrodynamic Stabilization and Control of Long Orbiting Tethers, Interim Report for Sep 1979—Feb 1981, Smithsonian Astrophysical Observatory., March 1981.
2. K. R Kroll, "Tethered Propellant Resupply Technique for Space Stations," IAF-84-442, presented at the 35th LAF Congress, Lausanne Switzerland, 1984.

Gravity Gradient Effects

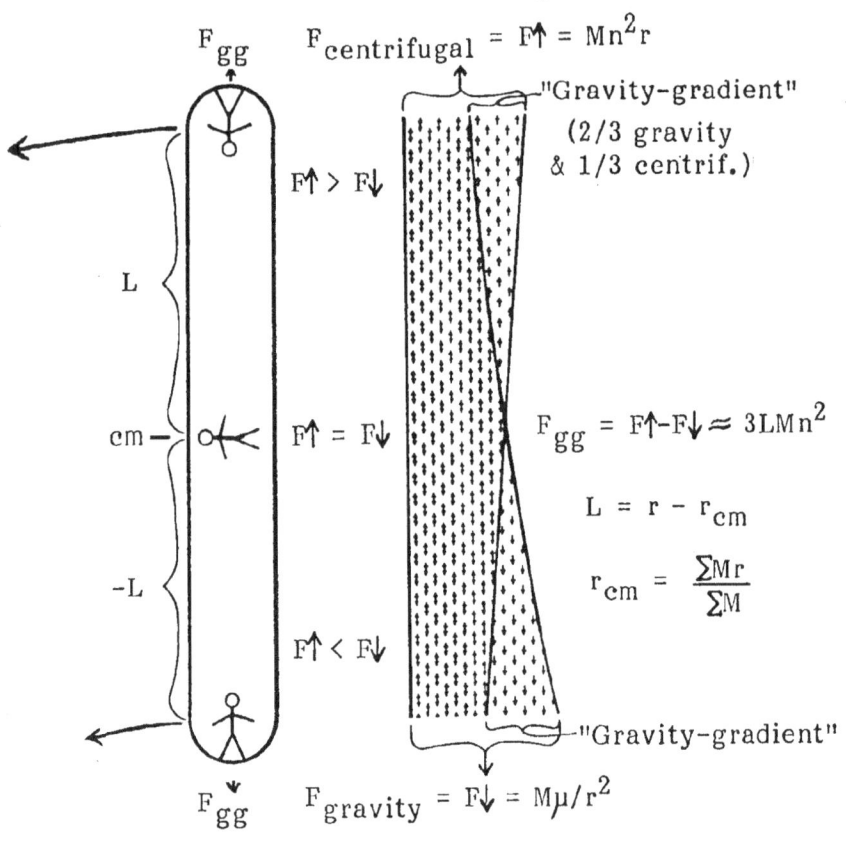

$F_{centrifugal} = F\uparrow = Mn^2r$

"Gravity-gradient"
(2/3 gravity & 1/3 centrif.)

$F\uparrow > F\downarrow$

$F\uparrow = F\downarrow$

$F_{gg} = F\uparrow - F\downarrow \approx 3LMn^2$

$L = r - r_{cm}$

$r_{cm} = \dfrac{\sum Mr}{\sum M}$

$F\uparrow < F\downarrow$

"Gravity-gradient"

$F_{gravity} = F\downarrow = M\mu/r^2$

Origin of "Gravity-Gradient" Forces

$<10^{-7}$ gee over .5 m

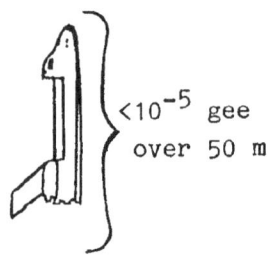

$<10^{-5}$ gee over 50 m

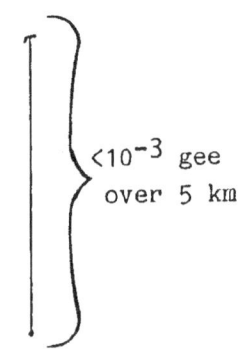

$<10^{-3}$ gee over 5 km

Magnitude of
Gravity Gradient
Effects in LEO

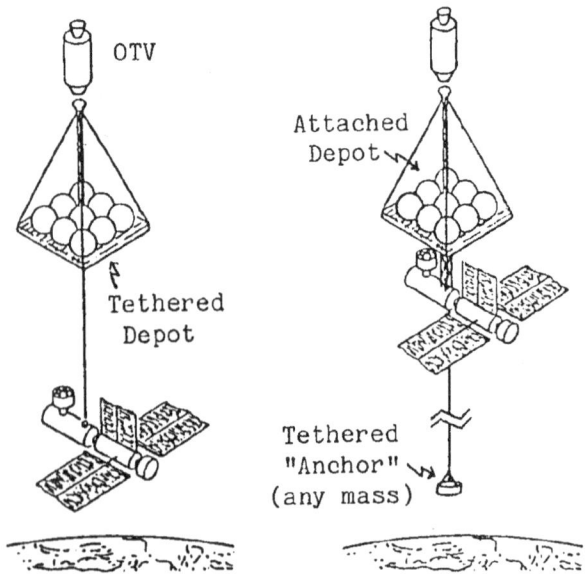

Two Propellant-Settling Options

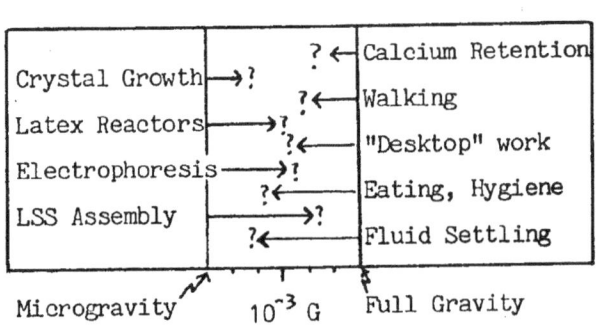

Potential Overlap of Regions for
Low-Gee & Gee-Dependent Operations

186

5.4.2 Dumbbell Libration in Circular Orbit

KEY POINTS   Libration periods are independent of length, but increase at large amplitude. Out-of-plane libration can be driven by weak forces that have a 2n component. Tethers can go slack if $\theta_{max}>65°$ or $\phi_{max}>60°$.

The two figures at right show the forces on a dumbbell in circular orbit which has been displaced from the vertical, and show the net torque on the dumbbell, returning it towards the vertical. The main difference between the two cases is that the centrifugal force vectors are radial in the in-plane case, and parallel in the out-of-plane case. This causes the net force in the out-of-plane case to have a smaller axial component and a larger restoring component, and is why ϕ-libration has a higher frequency than θ-libration.

Four aspects of this libration behavior deserve notice. First, the restoring forces grow with the tether length, so libration frequencies are independent of the tether length. Thus tether systems tend to librate "solidly", like a dumbbell, rather than with the tether trying to swing faster than the end-masses as can be seen in the chain of a child's swing. (This does not hold for very long tethers, since the gravity gradient itself varies.) For low orbits around any of the inner planets or the moon, libration periods are roughly an hour.

Second, tethered masses would be in free-fall except for the tether, so the sensed acceleration is always along the tether (as shown by the stick-figures). Third, the axial force can become negative, for $\phi>60°$ or near the ends of retrograde in-plane librations >65.9°. This may cause problems unless the tether is released, or retrieved at an adequate rate to prevent slackness.

NOTES   And fourth, although θ-libration is not close to resonance with any significant driving force, ϕ-libration is in resonance with several, such as out-of-plane components of aerodynamic forces (in non-equatorial orbits that see different air density in northward and southward passes) or electrodynamic forces (if tether currents varying at the orbital frequency are used). The frequency droop at large amplitudes (shown at right) sets a finite limit to the effects of weak but persistent forces, but this limit is quite high in most cases.

The equations given at right are for an essentially one-dimensional structure, with one principal moment of inertia far smaller than the other two: A<<B<C. If A is comparable to B & C, then the θ-restoring force shrinks with (B-A)/C, and the θ-libration frequency by Sqrt((B-A)/C). Another limitation is that a coupling between ϕ & θ behavior (see ref. 1) has been left out. This coupling is caused by the variation of end-mass altitudes twice in each ϕ-libration. This induces Coriolis accelerations that affect θ. This coupling is often unimportant, since 4n is far from resonance with 1.73n.

Libration is referenced to the local vertical, and when a dumbbell is in an eccentric orbit, variations in the orbital rate cause librations which in turn exert periodic torques on an initially uniformly-rotating object. In highly eccentric orbits this can soon induce tumbling.[2]

REFERENCES   1. D. Arnold, "General Equations of Motion," Appendix A of Investigation of Electrodynamic Stabilization and Control of Long Orbiting Tethers, Interim Report for Sep 1979—Feb 1981, Smithsonian Astrophysical. Observatory., March 1981.
2. P.A. Swan, "Dynamics & Control of Tethers in Elliptical Orbits," IAF-84-361, presented at the 35th IAF Congress, Lausanne, Switzerland, October 1984.

Dumbbell Libration in Circular Orbit

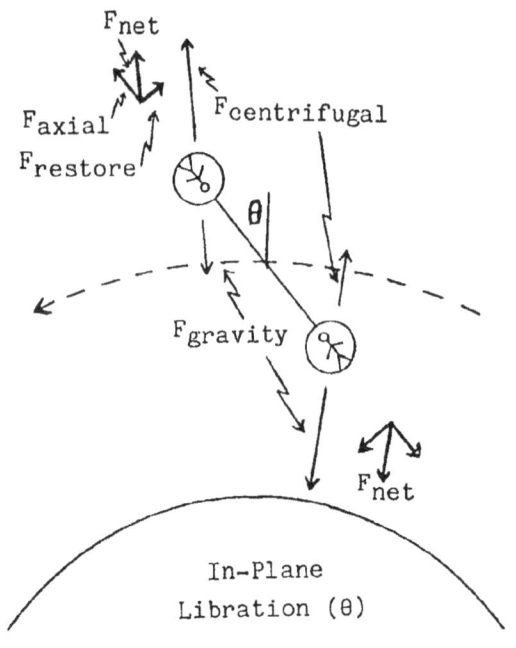

In-Plane
Libration (θ)

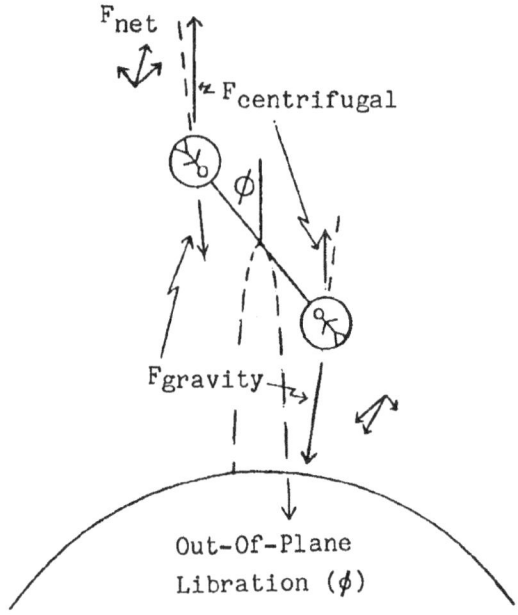

Out-Of-Plane
Libration (ϕ)

$$\ddot{\theta} \approx -3n^2\sin\theta\,\cos\theta = -1.5n^2\sin(2\theta)$$
$$\dot{\theta} \approx \pm\sqrt{3}\,n\sqrt{\sin^2\theta_{max}-\sin^2\theta}$$
$$(\dot{\theta} \approx \pm\sqrt{3}\,n\,\sin\theta_{max} \quad \text{when } \theta=0)$$
$$n_\theta \approx n\sqrt{3\cos\theta_{max}}$$

$$\ddot{\phi} \approx -4n^2\sin\phi\,\cos\phi = -2n^2\sin(2\phi)$$
$$\dot{\phi} \approx \pm2n\sqrt{\sin^2\phi_{max}-\sin^2\phi}$$
$$(\dot{\phi} \approx \pm2n\,\sin\phi_{max} \quad \text{when } \phi=0)$$
$$n_\phi \approx 2n\sqrt{\cos\phi_{max}}$$

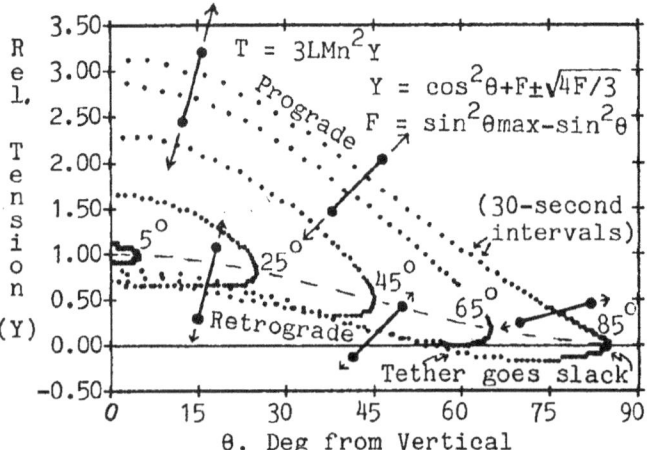

Tension Variations in Librating Dumbbells
(compared to tension in hanging dumbbells)

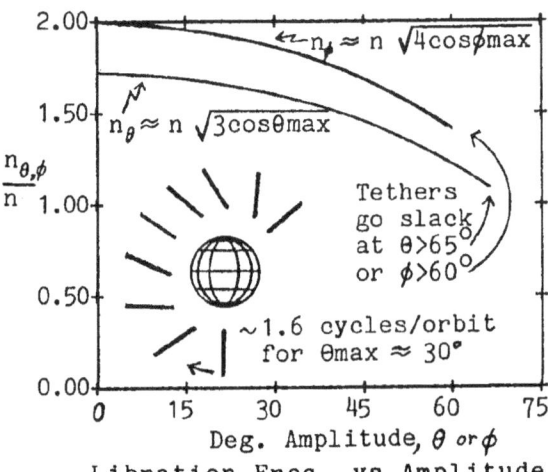

Libration Freq. vs Amplitude

188

5.4.3 Tether Control Strategies

KEY POINTS   Open-loop control is adequate for deployment; full retrieval requires feedback Tension laws can control θ & ϕ-libration plus tether oscillations. Many other options exist for libration, oscillation, & final retrieval control.

The table at right shows half a dozen distinct ways in which one or more aspects of tethered system behavior can be controlled. In general, anything which can affect system behavior (and possibly cause control problems) can be part of the solution, if it itself can be controlled without introducing other problems.

Thus, for example, stiff tethers have sometimes been considered undesirable, because the stiffness competes with the weak gravity-gradient forces near the end of retrieval. However, if the final section of tether is stiff AND nearly straight when stress-free (rather than pig-tail shaped), then "springy beam" control laws using a steerable boom tip might supplement or replace other laws near the end of retrieval. A movable boom has much the same effect as a stiff tether & steerable boom tip, since it allows the force vector to be adjusted.

NOTES   The basic concepts behind tension-control laws are shown at right. Libration damping is done by paying out tether when the tension is greater than usual and retrieving it at other times. This absorbs energy from the libration. As shown on the previous page, in-plane libration causes large variations in tension (due to the Coriolis effect), so "yoyo" maneuvers can damp in-plane librations quickly. Such yoyo maneuvers can be superimposed on deployment and retrieval, to allow large length changes (>4:1) plus large in-plane libration damping (or initiation) in less than one orbit, as proposed by Swet.[1]

Retrieval laws developed for the TSS require more time than Ref. 1, because they also include damping of out-of-plane libration built up during station keeping. Rupp developed the first TSS control law in 1975;[2] much of the work since then is reviewed in (3). Recent TSS control concepts combine tension and thrust control laws, with pure tension control serving as a backup in case of thruster failure.[4] Axial thrusters raise tether tension when the tether is short, while others control yaw & damp out-of-plane libration to allow faster retrieval.

A novel concept which in essence eliminates the final low-tension phase of retrieval is to have the end mass climb up the tether.[5] Since the tether itself remains deployed, its contribution to gravity-gradient forces and stabilization remains. The practicality of this will vary with the application.

REFERENCES   1. C. J. Swet, "Method for Deploying and Stabilizing Orbiting Structures",U.S. Patent #3,532,298, October 6, 1970.
2. C. C. Rupp, A Tether Tension Control Law for Tether Subsatellites Deployed Along Local Vertical, NASA TM X-64963, MSFC, September 1, 1975.
3. V. J. Modi, Geng Chang-Fu, A.R Misra, and Da Ming Xu, "On the Control of the Space Shuttle Based Tethered Systems," Acta Astronautica, Vol. 9, No. 6-7, pp. 437-443, 1982.
4. A. K. Banerjee and T.R. Kane, "Tethered Satellite Retrieval with Thruster Augmented Control," AIAA 82-1-21, presented at the AIAA/AAS Astrodynamics Conference, San Diego, Calif., 1982.
5. T. R. Kane, "A New Method for the Retrieval of the Shuttle-Based Tethered Satellite," J. of the Astronaut. Sci., Vol 32, No. 3, July-Sept. 1984.

Tether Control Strategies

EFFECTIVENESS OF VARIOUS CONTROL CONCEPTS

APPLICATION CONTROL OUTPUT	Libration		Tether Oscillations		Endmass Attitude Osc.	
	In-plane	Out-of-plane	Longitudinal	Transverse	Pitch & Roll	Yaw
Tension	Strong	Weak	Strong	Strong	Strong	None
	(Note: tension control is weak when tether is short)					
El. Thrust	Only if M1 ≠ M2		None	Only odd harmonics	None	None
Thruster	Strong, but costly if prolonged			None	Strong, but costly if prolonged	
Movable mass	Good w/short tether		Possible but awkward		None	None
Stiff tether, Movable boom	Strong if tether is very short; weak otherwise					
Aerodynamic	High drag—use only if low altitude needed for other reasons.					

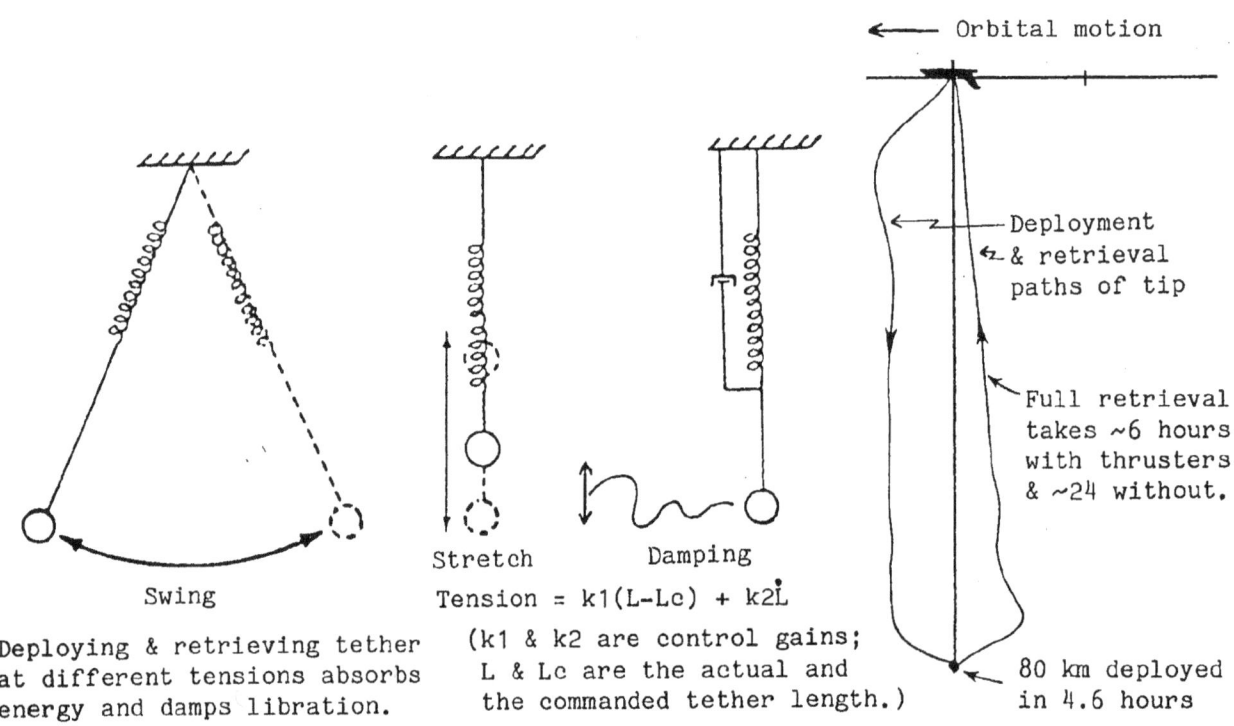

Swing

Deploying & retrieving tether at different tensions absorbs energy and damps libration.

Stretch Damping

Tension = k1(L-Lc) + k2L̇

(k1 & k2 are control gains;
L & Lc are the actual and
the commanded tether length.)

Orbital motion

Deployment & retrieval paths of tip

Full retrieval takes ~6 hours with thrusters & ~24 without.

80 km deployed in 4.6 hours

TENSION CONTROL FOR LIBRATION DAMPING... AND DEPLOYMENT/RETRIEVAL

190

5.4.4 Momentum Transfer Without Release

KEY POINTS   Tethers merely redistribute angular momentum; they do not create it. Changes in tether length, libration, and spin all redistribute momentum. Momentum transfer out-of-plane or in deep space is possible but awkward.

The two figures at right show two different tether deployment (and retrieval) techniques. In both cases, the initial deployment (which is not shown) is done with RCS burns or a long boom. In the case at left, the tether is paid out under tension slightly less than the equilibrium tension level for that tether length. The tether is slightly tilted away from the vertical during deployment, and librates slightly after deployment is complete.

In the other case, after the initial near-vertical separation (to about 2% of the full tether length), the two end masses are allowed to drift apart in near-free-fall, with very low but controlled tension on the tether. Just under one orbit later, the tether is almost all deployed and the range rate decreases to a minimum (due to orbital mechanics). RCS burns or tether braking are used to cushion the end of deployment and prevent end mass recoil. Then the tether system begins a large-amplitude prograde swing towards the vertical.

NOTES   In both cases, the angular momentum transferred from one mass to the other is simply, as stated in the box, the integral over time of the radius times the horizontal component of tether tension. In one case, transfer occurs mainly during deployment; in the other, mainly during the libration after deployment. In each case, momentum transfer is greatest when the tether is vertical, since the horizontal component of tether tension changes sign then.

An intermediate strategy—deployment under moderate tension—has also been investigated. However, this technique results in very high deployment velocities and large rotating masses. It also requires powerful brakes and a more massive tether than required with the other two techniques.

As discussed under Tether Control Strategies, changing a tether's length in resonance with variations in tether tension allows pumping or damping of libration or even spin. Due to Coriolis forces, in-plane libration and spin cause far larger tension variations than out-of-plane libration or spin, so in-plane behavior is far easier to adjust than out-of-plane behavior. Neglecting any parasitic losses in tether hysteresis & the reel motor, the net energy needed to induce a given libration or spin is simply the system's spin kinetic energy relative to the local vertical, when the system passes through the vertical.

Two momentum transfer techniques which appear applicable for in-plane, out-of-plane, or deep-space use are shown at right. The winching operation can use lighter tethers than other tethered-momentum-transfer techniques, but requires a very powerful deployer motor. The tangential ΔV simply prevents a collision.

The spin-up operation (proposed by Harris Mayer) is similar to the winching operation. It uses a larger tangential ΔV, a tether with straight and tapered sections, and a small motor. Retrieval speeds up the spin by a factor of $1/L^2$. Surprisingly, the long tapered section of tether can be less than half as massive as the short straight section that remains deployed after spin-up.

REFERENCES   1. J. Tschirgi, "Tether-Deployed SSUS-A, Report on NASA Contract NAS8-32842, McDonnell Douglas, April 1984.

Momentum Transfer

Retrieve   Deploy

High Tension
During Swing

Note tilt

$$\Delta(MVr) = \int rT\sin\theta\,\delta t$$

Low
Tether
Tension

Momentum Transfer During
Deployment & Retrieval

Momentum Transfer During Libration
(after low-tension deployment)

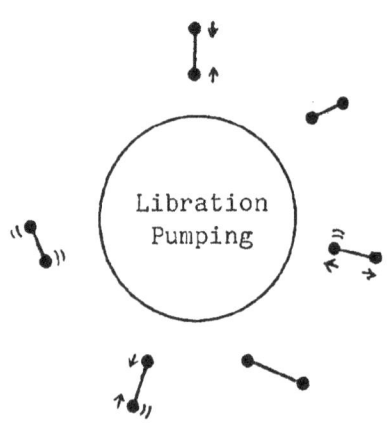

Libration
Pumping

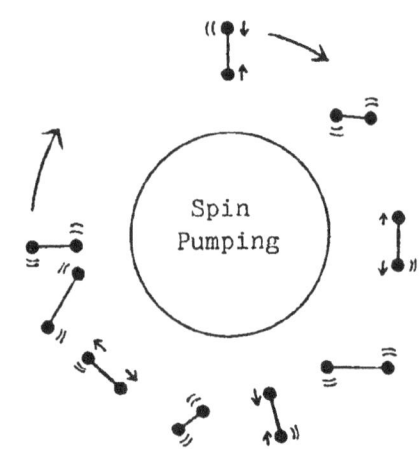

Spin
Pumping

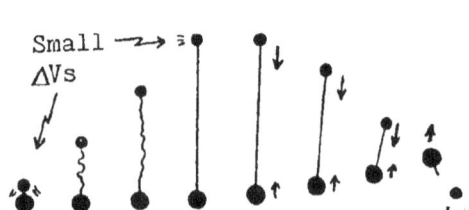

Small
ΔVs

Deployment Followed by Winching
(in orbit or in deep space)

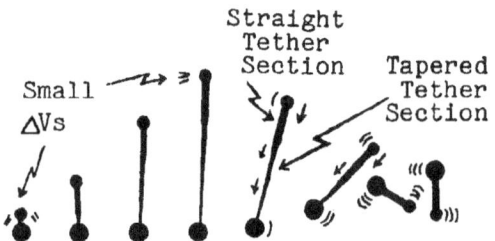

Straight
Tether
Section

Tapered
Tether
Section

Small
ΔVs

One Spin-Up Technique
For Use in Deep Space

192

5.4.5 Orbit Transfer by Release or Capture

KEY POINTS   The achievable orbit change scales with the tether length (as long as $\Delta r \ll r$). Retrograde-libration releases are inefficient, but allow concentric orbits. Apogee & perigee boosts have different values in different applications. Tethered capture can be seen as a time-reversal of a tether release operation.

The figures to the right show the size of the orbit changes caused by various tether operations. When released from a vertical tether, the end masses are obviously one tether length apart in altitude. The altitude difference 1/2 orbit later, Δr_{π}, varies with the operation but is usually far larger. The linear relationship shown becomes inaccurate when Δr approaches r. Tethered plane changes are generally limited to a few degrees and are not covered here.

Tether release leaves the center-of-mass radius at each phase angle roughly unchanged: if the upper mass is heavier, then it will rise less than the lower mass falls, and vice-versa. Note that the libration amplitude, θ_{max}, is taken as positive during prograde libration and negative during retrograde libration. Hence retrograde libration results in $\Delta r < 7L$. In particular, the pre-release & post-release orbits will all be concentric if $\theta_{max} = -60°$. But since methods of causing -60° librations usually involve +60° librations (which allow much larger boosts by the same tether), prograde releases may usually be preferable unless concentric orbits are needed or other constraints enter in.

NOTES   The relative tether length, mass, peak tension, and energy absorbed by the deployer brake during deployment as a function of (prograde) libration angle are all shown in the plot at right. Libration has a large effect on brake energy. This may be important when retrieval of a long tether is required, after release of a payload or after tethered-capture of a free-flying payload.

The double boost-to-escape operation at right was proposed by A. Cutler. It is shown simply as an example that even though momentum transfer is strictly a "zero sum game", a tethered release operation can be a "WIN-win game" (a large win & a small one). The small win on the deboost-end of the tether is due to the reduced gravity losses 1/2 orbit after release, which more than compensate for the deboost itself. Another example is that deboosting the shuttle from a space station can reduce both STS-deboost & station-reboost requirements.

Rendezvous of a spacecraft with the end of a tether may appear ambitious, but with precise relative-navigation data from GPS (the Global Positioning System) it may not be difficulty The relative trajectories required are simply a time-reversal of relative trajectories that occur after tether release. Approach to a hanging-tether rendezvous is shown at right. Prompt capture is needed with this technique: if capture is not achieved within a few minutes, one should shift to normal free-fall techniques. Tethered capture has large benefits in safety (remoteness) and operations (no plume impingement; large fuel savings). The main hazard is collision, due to undetected navigation or tether failure.

REFERENCES 1. G. Colombo, "Orbital Transfer & Release of Tethered Payloads," SAO report on NASA Contract NAS8-33691, March 1983.
2. W.D. Kelly, "Delivery and Disposal of a Space Shuttle External Tank to Low Earth Orbit," J. of the Astronaut. Sci., Vol. 32, No. 3, July-Sept 1984.
3. J.A. Carroll, "Tether-Mediated Rendezvous," report to Martin Marietta on Task 3 of contract RH3-393855, March 1984.
4. J.A. Carroll, "Tether Applications in Space Transportation, IAF 84-438, at the 35th IAF Congress, Oct 1984. To be published in ACTA ASTRONAUTICA.

Orbit Transfer by Tethered Release or Capture

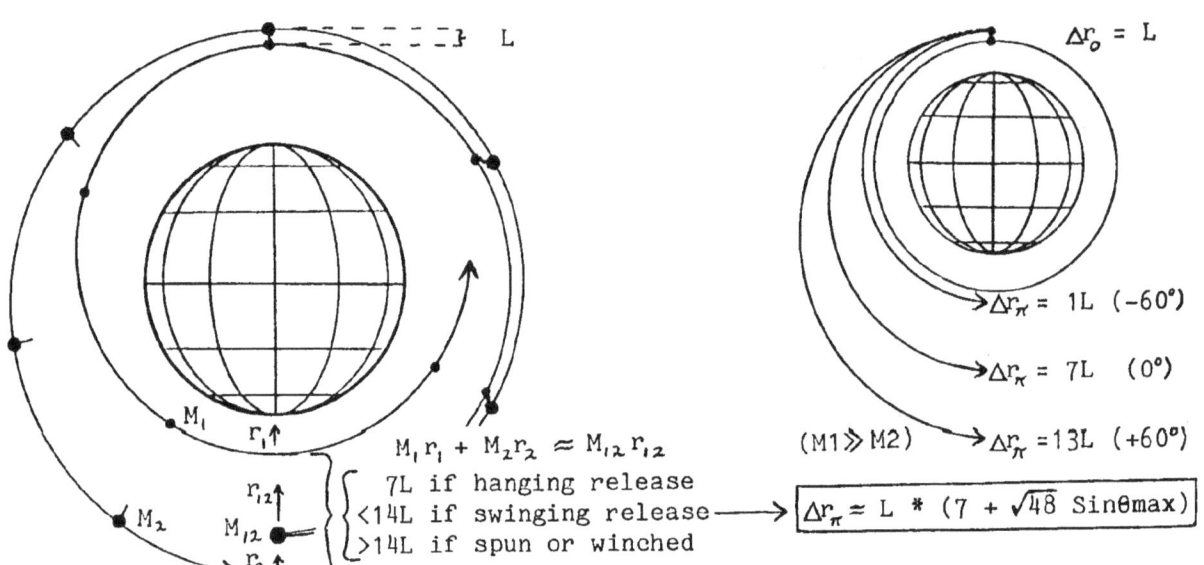

$$M_1 r_1 + M_2 r_2 \approx M_{12} r_{12}$$

$$\begin{cases} 7L \text{ if hanging release} \\ <14L \text{ if swinging release} \\ >14L \text{ if spun or winched} \end{cases} \longrightarrow$$

$$\boxed{\Delta r_\pi \simeq L * (7 + \sqrt{48}\ \text{Sin}\theta\text{max})}$$

$\Delta r_o = L$

$\Delta r_\pi = 1L \quad (-60°)$

$\Delta r_\pi = 7L \quad (0°)$

$(M1 \gg M2) \longrightarrow \Delta r_\pi = 13L \quad (+60°)$

Effects of Tether Deployment and Release

Effect of Libration on Boost
(release at middle of swing)

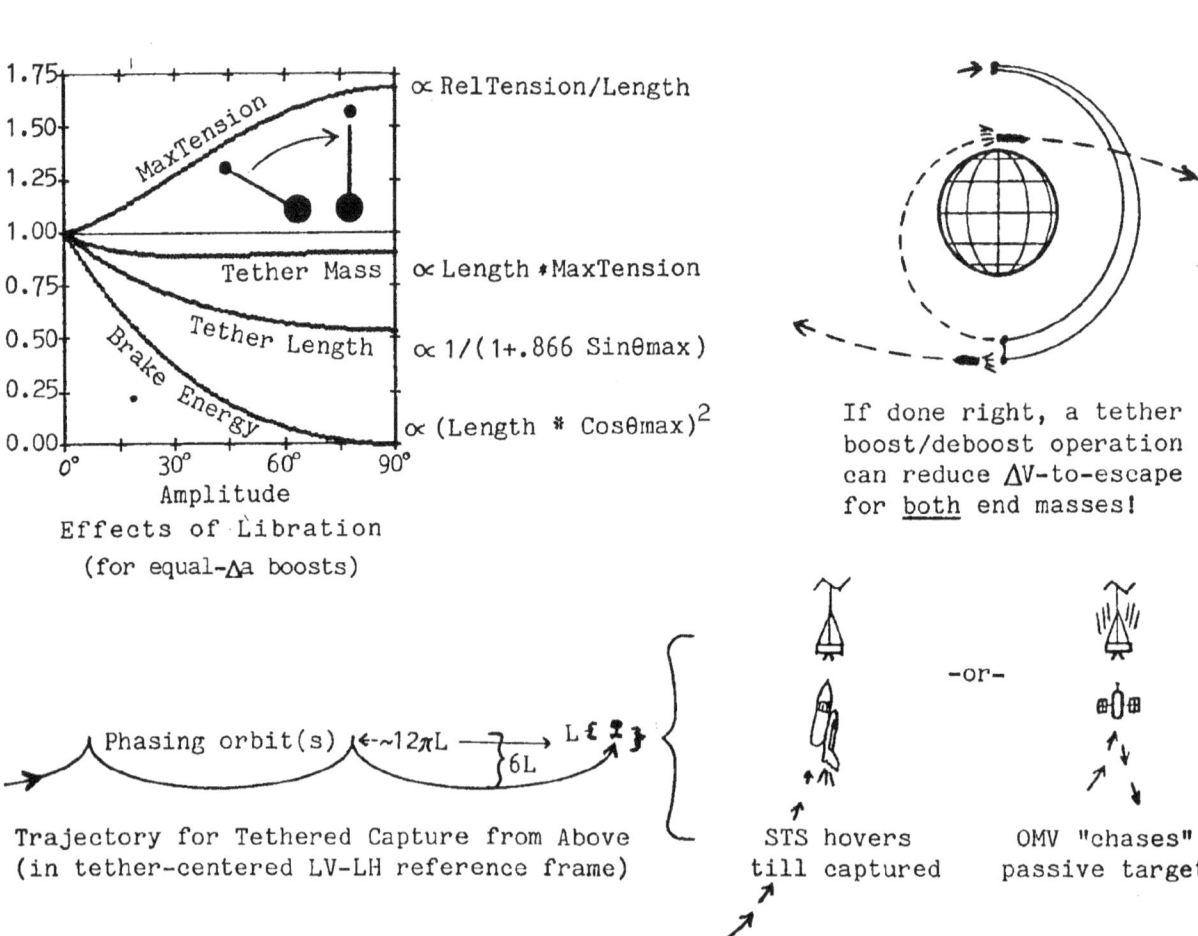

$\propto \text{RelTension/Length}$

$\propto \text{Length} * \text{MaxTension}$

$\propto 1/(1+.866\ \text{Sin}\theta\text{max})$

$\propto (\text{Length} * \text{Cos}\theta\text{max})^2$

Effects of Libration
(for equal-Δa boosts)

If done right, a tether
boost/deboost operation
can reduce ΔV-to-escape
for <u>both</u> end masses!

Trajectory for Tethered Capture from Above
(in tether-centered LV-LH reference frame)

STS hovers
till captured

OMV "chases"
passive target

194

5.4.6 Energy and Angular Momentum Balance

KEY POINTS   Tether operations cause higher-order repartitions of energy & angular momentum. First-order approximations that neglect these effects may cause large errors. Extremely long systems have strange properties such as positive orbital energy.

The question and answer at right are deceptively simple. The extent to which this is so, and the bizarre effects which occur in extreme cases, can be seen in the 3 graphs at right. At top, deploying & retrieving two masses on a very long massless tether changes not only the top & bottom orbital radii but also that of the CM. In addition, the free-fall location drops below the CM. Other key parameter changes under the same conditions are plotted underneath.

Note that when the tether length exceeds about 30% of the original orbital radius, the entire system lies below the original altitude. Also, at a radius ratio near 1.95:1, the maximum tether length compatible with a circular orbit is reached. At greater lengths (and the initial amount of angular momentum), no circular orbit is possible at any altitude.

Tether retrieval at the maximum-length point can cause the system to either rise or drop, depending on the system state at that time. If it continues to drop, there is a rapid rise in tether tension, and the total work done by the deployer quickly becomes positive. This energy input eventually becomes large enough (at 2.89:1) to even make the total system energy positive. The system is unstable beyond this point: any small disturbance will grow and can cause the tether system to escape from the body it was orbiting. (See ref. 2.)

NOTES

The case shown is rather extreme: except for orbits around small bodies such as asteroids, tethers either will be far shorter than the orbital radius, or will greatly outweigh the end masses. Either change greatly reduces the size of the effects shown. The effects on arbitrary structures can be calculated using the equations listed at right, which are based on a generalization of the concept of "moments" of the vertical mass distribution. Changes in tether length or mass distribution leave h unchanged, so other parameters (including r_{cm}, n, and E) must change. (For short tethers, the changes scale roughly with the square of the system's radius of gyration.) In many cases different conditions are most easily compared by first finding the orbital radius that the system would have if its length were reduced to 0, r_{Lt} = 0.

The mechanism that repartitions energy and angular momentum is that length changes cause temporary system displacements from the vertical. This causes both torques and net tangential forces on the system, which can be seen by calculating the exact net forces and couples for a non-vertical dumbbell. The same effect occurs on a periodic basis with librating dumbbells, causing the orbital trajectory to depart slightly from an elliptical shape.

Other topics which are beyond the scope of this guidebook but whose existence should be noted are: eccentricity changes due to deployment, orbit changes due to resonant spin/orbit coupling, and effects of 2- & 3-dimensional structures.

REFERENCES   1. G. Colombo, M. Grossi, D. Arnold, & M. Martinez-Sanchez, "Orbital Transfer and Release of Tethered Payloads," continuation of NAS8-33691, final report for the period Sep 1979—Feb 1983, Smithsonian Astrophysical Observatory, March 1983. (In particular, see the table on page 21.)
2. D. Arnold, "Study of an Orbiting Tethered Dumbbell System Having Positive Orbital Energy," addendum to final report on NAS8-35497, SAO, Feb 1985.

Energy & Momentum Balance

Question: What are the sources of the dumbbell spin angular momentum and deployer brake energy?

Answer: Orbit changes which repartition h & E.

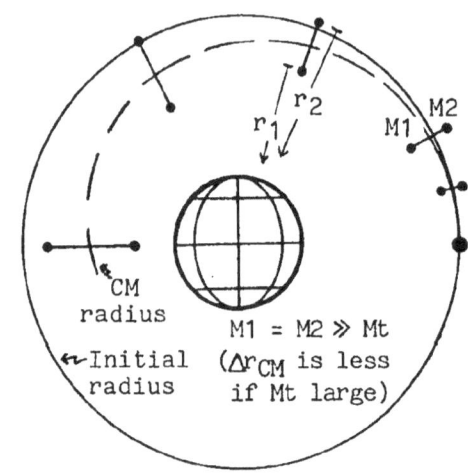

CM radius

Initial radius

M1 = M2 ≫ Mt
(Δr_{CM} is less if Mt large)

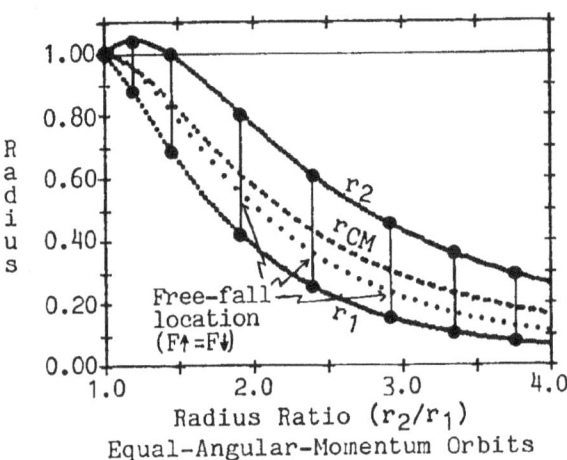

VERY-LONG-TETHER EFFECTS:

Radius

r_2
r_{CM}
Free-fall location ($F\uparrow = F\downarrow$)
r_1

Radius Ratio (r_2/r_1)
Equal-Angular-Momentum Orbits

For arbitrary nearly-one-dimensional vertical structures in circular orbit, analysis can be based on 5 "moments":

$$I_N = \sum M_i r_i^N \quad \text{(for N: -2..2)}$$

Each of these has physical meaning:

$$F_{grav} = \mu \, I_{-2}$$
$$E_{pot} = -\mu \, I_{-1}$$
$$Mass = I_0$$
$$F_{cen} = n^2 I_1$$
$$h_{tot} = n \, I_2$$
$$E_{kin} = .5 n^2 I_2$$

Some other useful equations include:

$$r_{cm} = I_1/I_0$$
$$n^2 = \mu \, I_{-2}/I_1$$
$$E = \mu((.5 I_{-2} * I_2/I_1) - I_{-1})$$
$$r_{(L_t=0)} = I_{-2}(I_2)^2/(I_1*(I_0)^2)$$

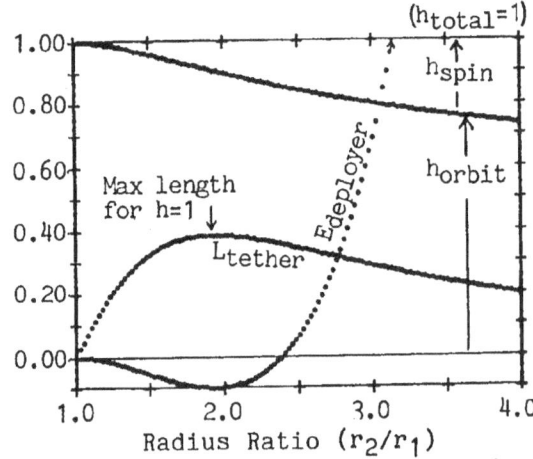

($h_{total}=1$)

h_{spin}
h_{orbit}

Max length for h=1
L_{tether}
$E_{deployer}$

Radius Ratio (r_2/r_1)
Angular Momentum Repartitioning, Tether Length, & Deployer Work

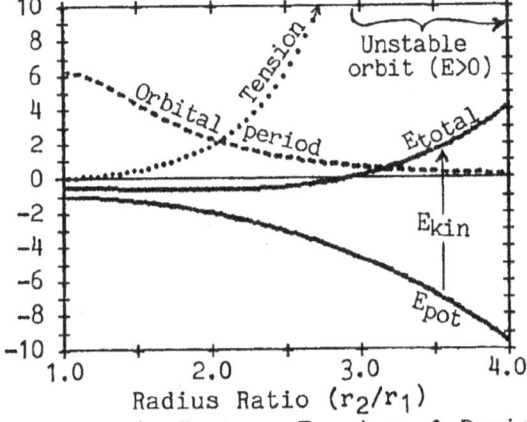

Tension
Orbital period
Unstable orbit (E>0)
E_{total}
E_{kin}
E_{pot}

Radius Ratio (r_2/r_1)
Changes in Energy, Tension, & Period

196

5.5 Tether Material Consideration
5.5.1 Tether Strength and Mass

KEY POINTS   Tether strength/weight ratio constrains performance in ambitious operations. Required tether mass is easily derivable from deltaV and payload mass.

Usable specific strength can be expressed in various ways. Three ways are shown at right. V_c, L_c, and L_{1g} are here defined in terms of a typical design stress (new/m^2) rather than the (higher) ultimate stress. Including the safety factor here streamlines the subsequent performance calculationsv Higher safety factors are needed with non-metals than with metals since non-metals are often more variable in their properties, brittle, abrasion-sensitive, and/or creep-sensitive. A safety factor of 4 (based on short-term fiber strength) is typical for Kevlar, but the most appropriate safety factor will vary with the applicatiol;

The "characteristic velocity," V_c, is the most useful parameter in tetherboost calculations, because the tether mass can be calculated directly from $\Delta V/V_c$, independently of the orbit, and nearly independently of the operation. The table at the bottom, which lists tether/rocket combinations that have the lowest lifesycle mass requirements, holds whenever $k_{Vc}=1$ km/sec & $I_{sp}=350$ sec.

The characteristic length L_c is useful in hanging-tether calculations. It varies with the orbital rate n. (The simple calculation given assumes L<<r; if this is not true, l/r effects enter in, and calculations such as those used in refs 3-5 must be used.) The safe 1-gee length L_{1g} is mainly useful in terrestrial applications, but is included since specific strength is often quoted this way. (Note that V_c and L_c vary with Sqrt(strength), and L_{1g} directly with strength).

NOTES   The specific modulus is of interest because it determines the speed of sound in the tether (C=the speed of longitudinal waves), the strain under design load ($\Delta L/L=\{V_c^2/C\}^2$), & the recoil speed after failure under design load (= V_c^2/C).

Tether mass calculations are best done by considering each end of the tether separately. If $M_{p1}>>M_{p2}$, then M_{t1} can be neglected in preliminary calculations.

Du Pont's Kevlar is the highest-specifiestrength fiber commercially available. Current RND efforts on high-performance polymers indicate that polyester can exhibit nearly twice the strength of Kevlar.[2] Two fiber producers have already announced plans to produce polymers with twice the specific strength of Kevlar.

In the long run, the potential may be greater with inorganic fibers like SiC & graphite. Refs. 3-5 focus on the requirements of "space elevators." They discuss laboratory tests of single-crystal fibers and suggest that 10-fold improvements in specific strength (or 3-fold in V_c & L_c) are conceivable.

REFERENCES 1. "Characteristics and Uses of Kevlar 49 Aramid High Modulus Organic Fiber" available from Du Pont's Textile Fibers Department, 1978.
2. G. Graff, "Superstrong Plastics Challenge Metals," High Technology magazine, February 1985, pp. 62-63.
3 J. Isaacs, H. Bradner, G. Backus, and A.Vine, "Satellite Elongation into a True "Skyhook"; a letter to Science, Vol. 151, pp. 682-683, Feb 11, 1966.
4. J. Pearson, "The Orbital Tower: a Spacecraft Launcher Using the Earth's Rotational Energy," Acta Astronautica, Vol.2, pp. 785-799, Pergamon, 1975.
5. H. Moravec, "A Non-Synchronous Orbital Skyhook," J. of the Astronautical Sciences, Vol. YXV, No. 4, pp. 307-322, Oct-Dec 1977.

Specific Strength and Required Tether Mass

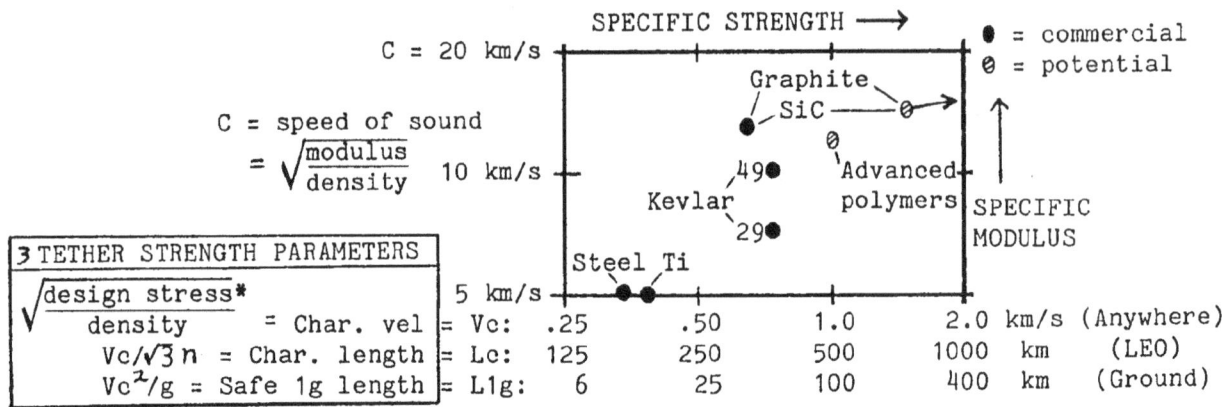

SPECIFIC STRENGTH ⟶   ● = commercial
                          ∅ = potential

C = 20 km/s

C = speed of sound
  = $\sqrt{\dfrac{\text{modulus}}{\text{density}}}$ 10 km/s

Graphite
  SiC
                49●   Advanced polymers
Kevlar
          29●   SPECIFIC MODULUS

3 TETHER STRENGTH PARAMETERS

$\sqrt{\dfrac{\text{design stress}^*}{\text{density}}}$ = Char. vel = Vc: .25   .50   1.0   2.0 km/s (Anywhere)

Vc/$\sqrt{3}\,n$ = Char. length = Lc: 125   250   500   1000 km (LEO)

Vc^2/g = Safe 1g length = L1g: 6   25   100   400 km (Ground)

Steel Ti   5 km/s

* Design stress is assumed to be 1/2 the ultimate strength for metals and 1/4 the short-term individual fiber strength for other materials.

SPECIFIC STRENGTH & MODULUS OF SEVERAL TETHER MATERIALS

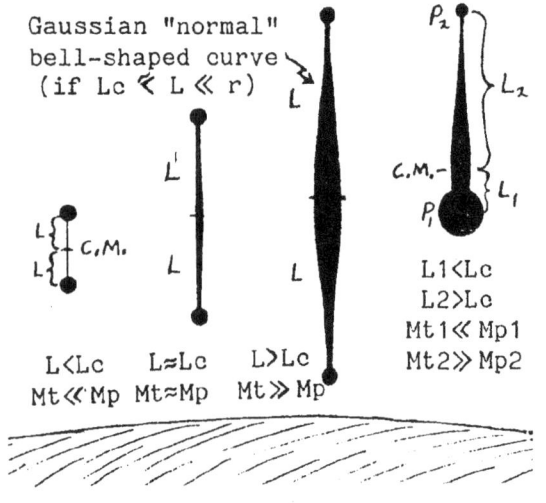

Gaussian "normal" bell-shaped curve (if Lc ≪ L ≪ r)

L1<Lc
L2>Lc
Mt1 ≪ Mp1
Mt2 ≫ Mp2

L<Lc   L≈Lc   L>Lc
Mt ≪ Mp   Mt ≈ Mp   Mt ≫ Mp

Tether Length & Required Mass

$\dfrac{Mt}{Mt+Mp}$

Untapered

Tapered

$\dfrac{Mt}{Mp} \to \sqrt{\dfrac{\pi}{2}}\,Xe^{x^2}$ for X>1

$\dfrac{Mt}{Mp} \to X^2$ for X<1

X (= ΔV/kVc, or L/Lc)**#**
Required Tether Mass (Mt)

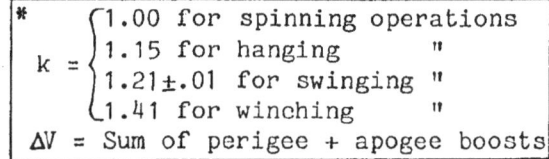

* k = { 1.00 for spinning operations
             1.15 for hanging   "
             1.21±.01 for swinging "
             1.41 for winching   "

ΔV = Sum of perigee + apogee boosts

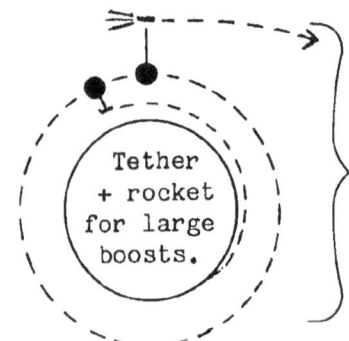

Tether + rocket for large boosts.

Expected # of uses	1	10	100	1000
Best tether ΔV, km/s	.14	.9	1.8	2.6
Required Mt/Mp	.02	1	11	95

(For kVc = 1 km/s and rocket Isp = 350 seconds; marginal deployer & dry rocket masses neglected.)

Best Tether ΔV for Combined Tether/Rocket Boosts

5.5.2 Tether Impact Hazards

KEY POINTS Micrometeoroids can sever thin tethers & damage tether protection/insulation. Orbiting debris (or other tethers) can sever tethers of any diameter. Debris could impact an Earth-based "Space Elevator" over once per year.

Sporadic micrometeoroids are usually assumed to have an typical density of about .5 and a typical impact velocity in LEO of approximately 20 km/sec.[1] At impact speeds above the speed of sound, solids become compressible and the impact shock wave has effects like those of an explosion. For this reason, the risk curve assumes that if the EDGE of an adequately large meteoroid comes close enough to the center of the tether (within 45° or .35 D_t), failure will result.

Experiments done by Martin Marietta on TSS candidate materials have used glass projectiles fired at 6.5 km/sec, below the (axial) speed of sound in Kevlar. Two damaged tethers from those tests are shown at right. The scaling law used ($\rho^{0.5}V^{0.67}$) indicates that this is representative of orbital conditions, but that law (used for impacts on sheet metal) may not apply to braided fibers.

For tethers much thicker than 10 mm or so (depending on altitude), the risk does not go down much as D_t increases, because even though the micrometeoroid risk still decreases, the debris risk (which INCREASES slightly with D_t) begins to dominate. As with micrometeoroids, the tether is assumed to fail if any part of the debris passes within 0.35 D_t of the center of the tether.

NOTES The debris risk at a given altitude varies with the total debris width at that altitude. This was estimated from 1983 CLASSY radar cross-section (RCS) data, by simply assuming that W = Sqrt(RCS) and summing Sqrt(RCS) over all tracked objects in LEO.[6] This underestimates W for objects with appendages, and over-estimates it for non-librating elongated objects without appendages.

CLASSY RCS data are expected to be accurate for RCS > 7 m². The 700 objects with RCS > 7 m² account for 3 km of the total 5 km width, so errors with smaller objects are not critical. Small untracked objects may not add greatly to the total risk: 40,000 objects averaging 2 cm wide would increase the risk to a 1-cm tether by only 20%. W was assumed independent of altitude, so the distribution of risk with altitude could be estimated by simply scaling Figure 1 from Ref. 4.

As shown at right, debris impact with a space elevator could be expected more than once per year at current debris populations. The relative density at 0° latitude was estimated from data on pp. 162-163 of ref. 6.

Similar calculations can be made for two tethers in different orbits at the same altitude. If at least one is spinning or widely-librating, the mutual risks can exceed 0.1 cut/km·yr. This makes "tether traffic control" essential.

REFERENCES 1. Meteoroid Environment Model—1969 [Near Earth to Lunar Surface], NASA SP-8013, March 1969.
2. Meteoroid Environment Model—1970 [Interplanetary and Planetary], NASA SP-8038, October 1970.
3. Meteoroid Damage Assessment, NASA SP-8042, May 1970. (Shows impact effects)
4. D. J. Kessler, "Sources of Orbital Debris and the Projected Environment for Future Spacecraft", in J. of Spacecraft & Rockets, Vol 18 #4, Jul-Aug 1981.
5. D. J. Kessler, Orbital Debris Environment for Space Station, JSC-20001, 1984.
6. CLASSY Satellite Catalog Compilations as of 1 Jan 1983, NORAD/J5YS, 1983.

Impact Hazards for Tethers

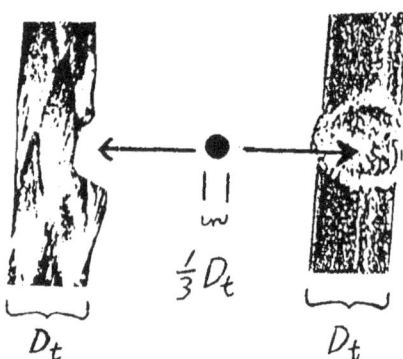

Braided
Kevlar,
grazed

Stainless
steel wire,
direct hit

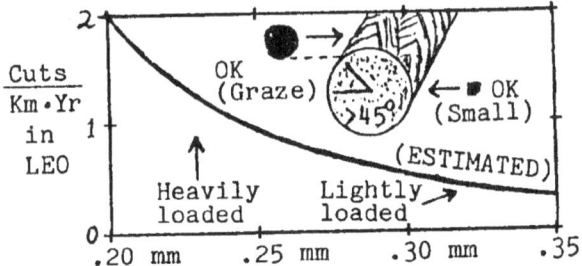

μMeteoroid Risks to a 1 mm Tether

For tethers with Dt > 1 (mm),
& Max non-fatal Dm = .25 Dt,

$$\frac{\mu m\ cuts}{Km \cdot Yr} \approx Dt^{-2.6}$$

Effective Width, W
(Any position between
the 2 extremes shown
cuts the tether.)

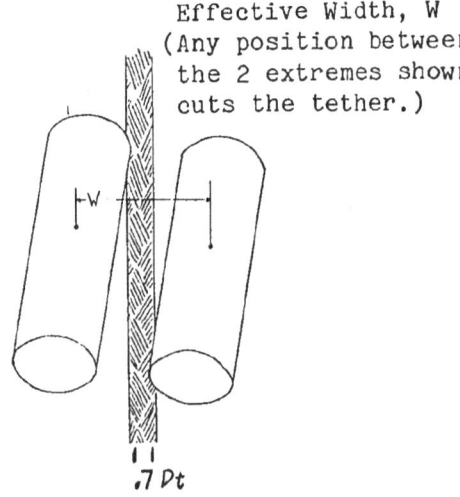

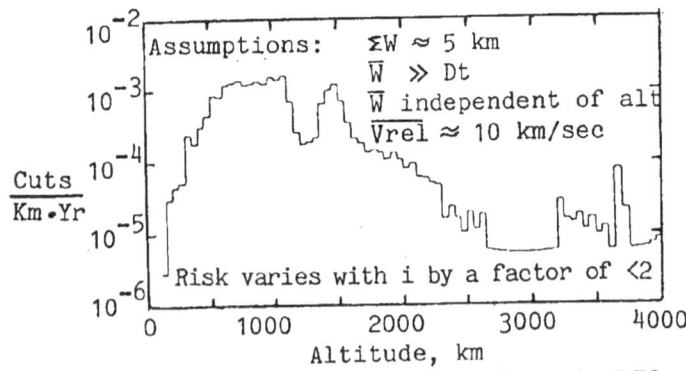

Debris Impact Rate on Tethers in LEO

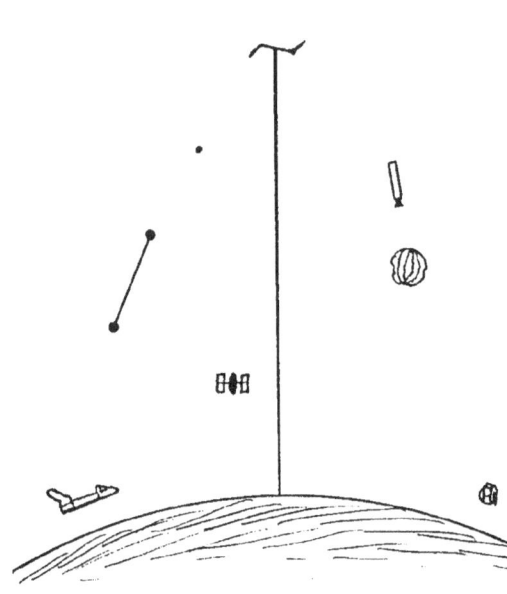

Debris Risk to the Lowest 4000 km
of an Earth-based Space Elevator:

$$Risk = \frac{\Sigma Width * \overline{V} * RelDensity\ at\ \lambda=0}{Earth\ "Surface\ Area"\ at\ Alt}$$

$$= \frac{\sim5\ km * \sim7.3\ km/sec * \sim.72}{4 * pi * Sqr(\sim7378\ km)}$$

$$\approx 3.9E\text{-}8/sec \approx \boxed{1.2\ cuts/year}$$

5.6 Electrodynamic Tethers
5.6.1 Interactions with Earth's Magnetic Field and Plasma
KEY POINTS Tether (& other) resistance can limit the output of electrodynamic tethers. Electron collection methods & effectiveness are important—and uncertain.

Since the publication of ref. 1, 20 years ago, electrodynamic tether proposals and concepts have been a frequent source of controversy, mainly in these areas:
1. What plasma instabilities can be excited by the current?
2. What is the current capacity of the plasma return loop?
3. What is the best way to collect electrons from the plasma?
The first Tethered Satellite mission may do much to answer these questions. The discussion below and graphics at right merely seek to introduce them.

The current flowing through an electrodynamic tether is returned in the surrounding plasma. This involves electron emission, conduction along geomagnetic field lines down to the lower ionosphere, cross-field conduction by collision with neutral atoms, and return along other field lines.

The tether current causes a force on the tether (and on the field) perpendicular to both the field and the tether (horizontal, if the tether is vertical). Motion of the tether through the geomagnetic field causes an EMF in the tether. This allows the tether to act as a generator, motor, or self-powered ultra-low-frequency broadcast antenna.[2] The motion also causes each region of plasma to experience only a short pulse of current, much as in a commutated motor.

NOTES Based on experience with charge neutralization of spacecraft in high orbit, it has been proposed that electrons be collected by emitting a neutral plasma from the end of the tether, to allow local cross-field conduction.[3] In GEO, the geomagnetic field traps a plasma in the vicinity of the spacecraft, and "escape" along field lines may not affect its utility. This may also hold in high-inclination orbits in LEO. But in low inclinations in LEO, any emitted plasma might be promptly wiped away by the rapid motion across field lines.

A passive collector such as a balloon has high aerodynamic drag, but a end-on sail can have an order of magnitude less drag. The electron-collection sketch at bottom right is based on a preliminary analysis by W. Thompson.[5] This analysis suggests that a current moderately higher than the electron thermal current ($=Ne * \sim 200$ km/sec) might be collected on a surface normal to the field. This is because collecting electrons requires that most ions be reflected away from the collection region as it moves forward. This pre-heats and densifies the plasma ahead of the collector. The voltage required for collection is just the voltage needed to repel most of the ions, about 12 V.

REFERENCES 1. S. D. Drell, H. M. Foley, & M. A. Ruderman, "Drag and Propulsion of Large Satellites in the Ionosphere: An Alfven Propulsion Engine in Space," J. Of Geophys. Res., Vol. 70, No. 13, pp. 3131-3145, July 1965.
2. M. Grossi, "A ULF Dipole Antenna on a Spaceborne Platform of the PPEPL Class," Report on NASA Contract NAS8-28203, May 1973.
3. R. D. Moore, "The Geomagnetic Thruster—A High Performance "Alfven Wave" Propulsion System Utilizing Plasma Contacts," ALGA Paper No. 66-257.
4. S. T. Wu, ed., University of Alabama at Huntsville/NASA Workshop on The Uses of a Tethered Satellite System, Summary Papers, Huntsville AL, 1978. See papers by M. Grossi et al, R. Williamson et al., and N. Stone.
5. W. Thompson, "Electrodynamic Properties of a Conducting Tether," Final Report to Martin Marietta Corp. on Task 4 of Contract RH3-393855, Dec. 1983.

201

Electrodynamic Tether Principles

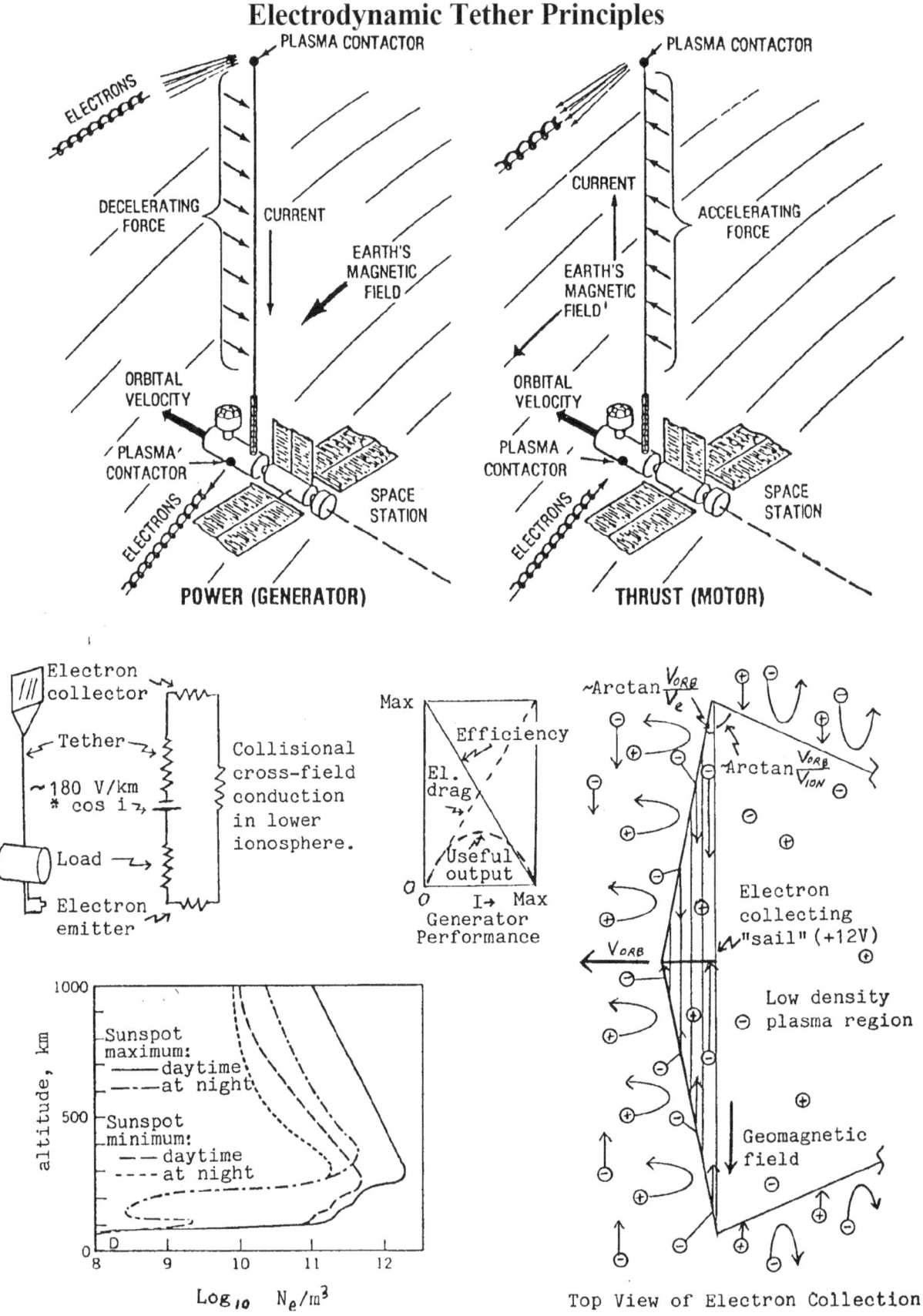

PLASMA CONTACTOR

ELECTRONS

DECELERATING FORCE

CURRENT

EARTH'S MAGNETIC FIELD

ORBITAL VELOCITY

PLASMA CONTACTOR

ELECTRONS

SPACE STATION

POWER (GENERATOR)

PLASMA CONTACTOR

CURRENT

ACCELERATING FORCE

EARTH'S MAGNETIC FIELD

ORBITAL VELOCITY

PLASMA CONTACTOR

ELECTRONS

SPACE STATION

THRUST (MOTOR)

Electron collector

Tether

~ 180 V/km * cos i

Load

Electron emitter

Collisional cross-field conduction in lower ionosphere.

Max

Efficiency

El. drag

Useful output

O   I→   Max

Generator Performance

~Arctan $\frac{V_{ORB}}{V_e}$

Arctan $\frac{V_{ORB}}{V_{ION}}$

Electron collecting "sail" (+12V)

Low density plasma region

V_{ORB}

Geomagnetic field

Top View of Electron Collection

altitude, km

1000

Sunspot maximum:
— daytime
—·— at night

Sunspot minimum:
— — daytime
····· at night

500

0

8   9   10   11   12

Log_{10} N_e/m^3

202

5.6.2 Electrodynamic Orbit Changes

KEY POINTS   Electrodynamic tether use will affect the orbit—whether desired or not. Station keeping and/or large orbit changes without propellant use are possible.

The offset dipole approximation shown at right is only a first approximation to the geomagnetic field: harmonic analyses of the field give higher-order coefficients up to 20% as large as the fundamental term. Ref. 1 contains computerized models suitable for use in detailed electrodynamic studies.

The geomagnetic field weakens rapidly as one moves into higher orbits, and becomes seriously distorted by solar wind pressure beyond GEO. However, ohmic losses in a tether are already significant in LEO, so electrodynamic tethers are mainly useful in low orbits where such distortions are not significant.

As the earth rotates, the geomagnetic field generated within it rotates also, and the geomagnetic radius and latitude of a point in inertial space vary over the day. If a maneuvering strategy which repeats itself each orbit is used (necessary unless the spacecraft has large diurnal power storage capacity), then the average effect, as shown at right, will be a due east thrust vector.

Variations in geomagnetic latitude (and thus in B_h) cancel out variations in the component of flight motion perpendicular to the field, so these variations do not cause large voltage variations in high-inclination orbits. (Note that the relevant motion is motion relative to a rotating earth.) Out-of-plane libration, variations in geomagnetic radius, and diurnal variation of the "geomagnetic inclination" of an orbit can all cause voltage variations. Peak EMFs (which drive hardware design) may approach 400 V/km.

NOTES   However these variations need not affect the thrust much if a spacecraft has a variable-voltage power supply: neglecting variations in parasitic power, constant power investment in a circular orbit has to give constant in-plane thrust. The out-of-plane thrust is provided "free" (whether desired or not). Average voltage & thrust equations for vertical tethers are shown at right.

The table shows how to change all six orbital elements separately or together. Other strategies are also possible. Their effects can be calculated from the integrals listed. For orbits within 11° of polar or equatorial, diurnally-varying strategies become more desirable. Computing their effects requires using the varying geomagnetic inclination instead of i (& moving it inside the integral). Note that the "DC" orbit-boosting strategy also affects i. This can be canceled out by superimposing a -2 Cos(2φ) current on the DC current.

As discussed under Electrodynamic Libration Control Issues, eccentricity and apside changes can strongly stimulate φ-libration unless the spacecraft center of mass is near the center of the tether. Other maneuvers should not do this, but this should be checked using high-fidelity geomagnetic field models.

REFERENCES 1. E. G. Stassinopoulos & G. D. Mead, ALLMAG, GDALMG, LINMA:Computer Programs for Geomagnetic Field & Field-Line Calculations, Feb. 1972, NASA Goddard.
2. R. D. Moore, "The Geomagnetic Thruster—A High Performance "Alfven Wave" Propulsion System Utilizing Plasma Contacts," AIAA Paper No. 66-257.
3. H. Alfven, "Spacecraft Propulsion; New Methods," Science, Vol. 176, 14 Apr 1972, pp. 167-168.

Electrodynamic Orbit Changes

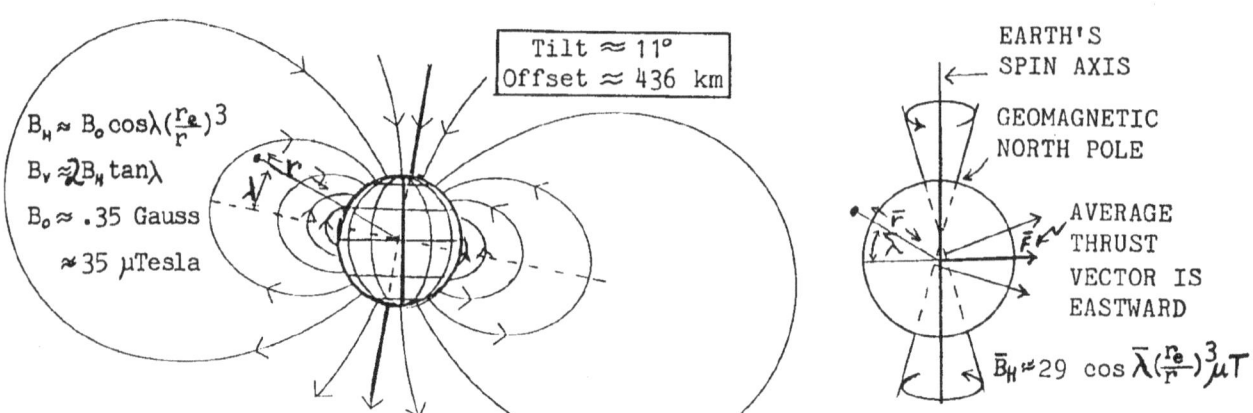

$$B_H \approx B_o \cos\lambda \left(\frac{r_e}{r}\right)^3$$

$$B_V \approx 2B_H \tan\lambda$$

$$B_o \approx .35 \text{ Gauss}$$

$$\approx 35 \ \mu\text{Tesla}$$

Tilt $\approx 11°$
Offset ≈ 436 km

OFFSET DIPOLE APPROXIMATION TO GEOMAGNETIC FIELD

EARTH'S SPIN AXIS

GEOMAGNETIC NORTH POLE

AVERAGE THRUST VECTOR IS EASTWARD

$$\bar{B}_H \approx 29 \cos\bar{\lambda}\left(\frac{r_e}{r}\right)^3 \mu T$$

EFFECT OF EARTH'S SPIN ON TIME-AVERAGED FIELD

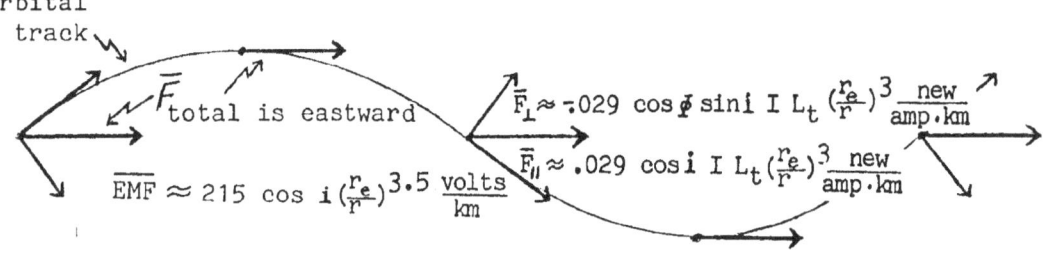

Orbital track

$\bar{F}_{total}$ is eastward

$$\bar{F}_\perp \approx .029 \cos\phi \sin i \ I \ L_t \left(\frac{r_e}{r}\right)^3 \frac{\text{new}}{\text{amp·km}}$$

$$\bar{F}_{||} \approx .029 \cos i \ I \ L_t \left(\frac{r_e}{r}\right)^3 \frac{\text{new}}{\text{amp·km}}$$

$$\overline{EMF} \approx 215 \cos i \left(\frac{r_e}{r}\right)^{3.5} \frac{\text{volts}}{\text{km}}$$

HOW TO CHANGE ORBITS USING AN ELECTRODYNAMIC TETHER

Element	Strategy	Thrust Vector	Effect
Semimajor axis	DC		$\Delta a \approx \cos(i)\frac{kl}{m}\int I\ dt$
Phase	Sawtooth		$\Delta M \approx \cos(i)\frac{1.5\,kln}{ma}\int I\ t\ dt$
Eccentricity	Cos(θ)		$\Delta e \approx \cos(i)\frac{kl}{ma}\int I \cos(\theta)\ dt$
Line of apsides	Sin(θ)		$\Delta w \approx \cos(i)\frac{kl}{mae}\int I \sin(\theta)\ dt$
Inclination	-Cos($2\bar{\phi}$)		$\Delta I \approx \frac{-kl}{2ma}\int I \sin(i)\cos^2(\bar{\phi})\ dt$
Ascending node	-Sin($2\bar{\phi}$)		$\Delta\Omega \approx \frac{-kl}{2ma}\int I \sin(\bar{\phi})\cos(\bar{\phi})\ dt$

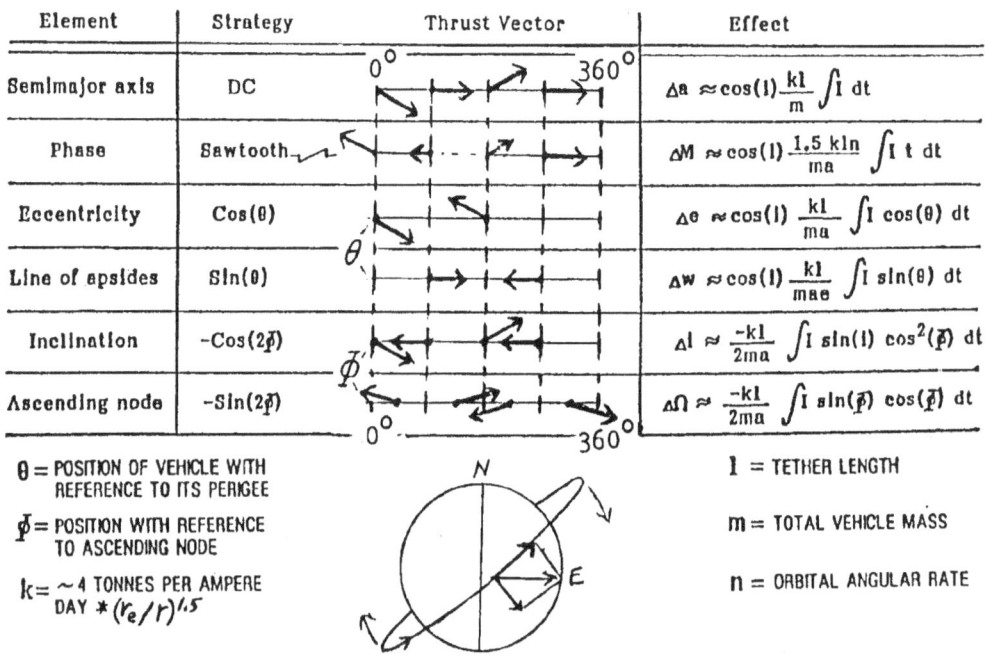

θ = POSITION OF VEHICLE WITH REFERENCE TO ITS PERIGEE

$\bar{\phi}$ = POSITION WITH REFERENCE TO ASCENDING NODE

k = ~4 TONNES PER AMPERE DAY * $(r_e/r)^{1.5}$

l = TETHER LENGTH

m = TOTAL VEHICLE MASS

n = ORBITAL ANGULAR RATE

5.6.3 Tether Shape and Libration Control

KEY POINTS Properly controlled AC components can be used to control θ and ϕ-libration. Solar-energy storage and e or ω changes strongly stimulate ϕ-libration. AC currents other than 1 & 3/orbit should not affect ϕ-libration much.

The maneuvering strategies on the previous page have assumed that electrodynamic tethers will stay vertical. However, as shown at right, the distributed force on the tether causes bowing, and that bowing is what allows net momentum transfer to the attached masses. Note that net momentum can be transferred to the system even if the wire is bowed the wrong way (as when the current is suddenly reversed); momentum transferred to the wire gets to the masses later.

This figure also illustrates two other issues:
1. Bowing of the tether causes it to cross fewer field lines.
2. Unequal end masses and uniform forces cause overall torques & tilting.

The bowing causes the tether to provide less thrust while dissipating the same parasitic power. The net force on the system is the same as if the tether were straight but in a slightly weaker magnetic field.

The torque on the system causes it to tilt away from the vertical, until the torque is balanced by gravity-gradient restoring torques. For a given system mass and power input, disturbing torques vary with L and restoring torques with L^2, so longer systems can tolerate higher power. The mass distribution also affects power-handling capability, as seen in the sequence at top right.

NOTES Modulating the tether current modulates any electrodynamic torques. Current modulation at 1.73 n can be used to control in-plane libration. Out-of-plane torques can also be modulated, but another control logic is required. This is because the once-per-orbit variation in out-of-plane thrust direction makes a current with frequency F (in cycles per orbit) cause out-of-plane forces and torques with frequencies of F-1 and F+1, as shown in the Fourier analysis at bottom right. Hence ϕ libration control (F=2) requires properly phased F=1 or F=3 currents. Higher frequencies can damp odd harmonics of any tether bowing oscillations. Control of both in- & out-of-plane oscillations may be possible since they have the same frequencies and thus require different currents.

Applications that require significant F=1 components for other reasons can cause problems. Four such strategies are shown at right. Sin & Cos controls allow adjustment of e or ω. The two "Sign of ..." laws allow constant power storage over 2/3 of each orbit and recovery the rest of the orbit. These laws would be useful for storing photovoltaic output for use during dark periods.

These strategies drive out-of-plane libration (unless the center of mass is at the center of the tether). The libration frequency decreases at large amplitudes, so if the system is not driven too strongly, it should settle into a finite-but-large-amplitude phase-locked loop. This may be unacceptable in some applications, due to resulting variations in gravity or tether EMF. In some cases, such as eccentricity changes, adding a F=3 component might cancel the undesired effect of an F=1 current while keeping the desired effect.

REFERENCE 1. G. Colombo, M. Grossi, M. Dobrowolny, and D. Arnold, Investigation of Electrodynamic Stabilization & Control of Long Orbiting Tethers, Interim Report on Contract NAS8-33691, March 1981, Smithsonian Astrophysical Observatory.

Electrodynamic Libration Control Issues

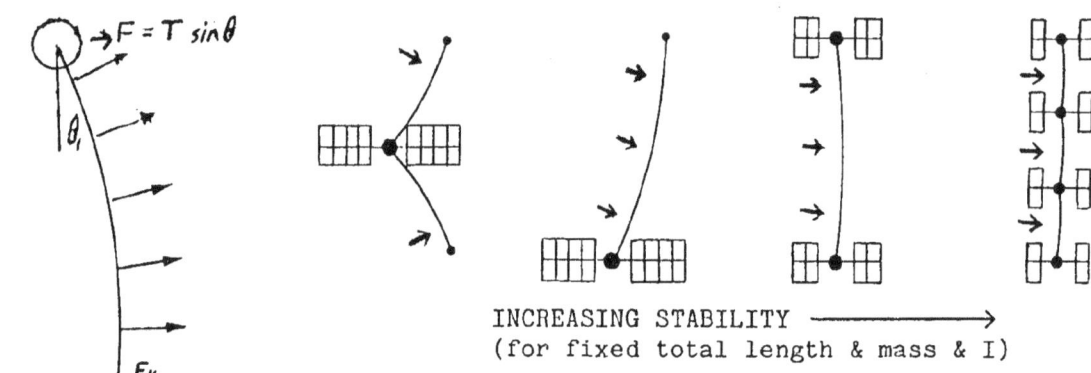

$F = T \sin\theta$

θ_1

$F_{\parallel}$

$F_{\perp}$

C.M. •

θ_2

INCREASING STABILITY ⟶
(for fixed total length & mass & I)

FOR CONTROL OF:	MODULATE I AT:
Out-of-plane libration*	1 n or 3 n
In-plane libration*	1.73 n
Tether oscillations	>5 n

* I or mass distribution must be lopsided

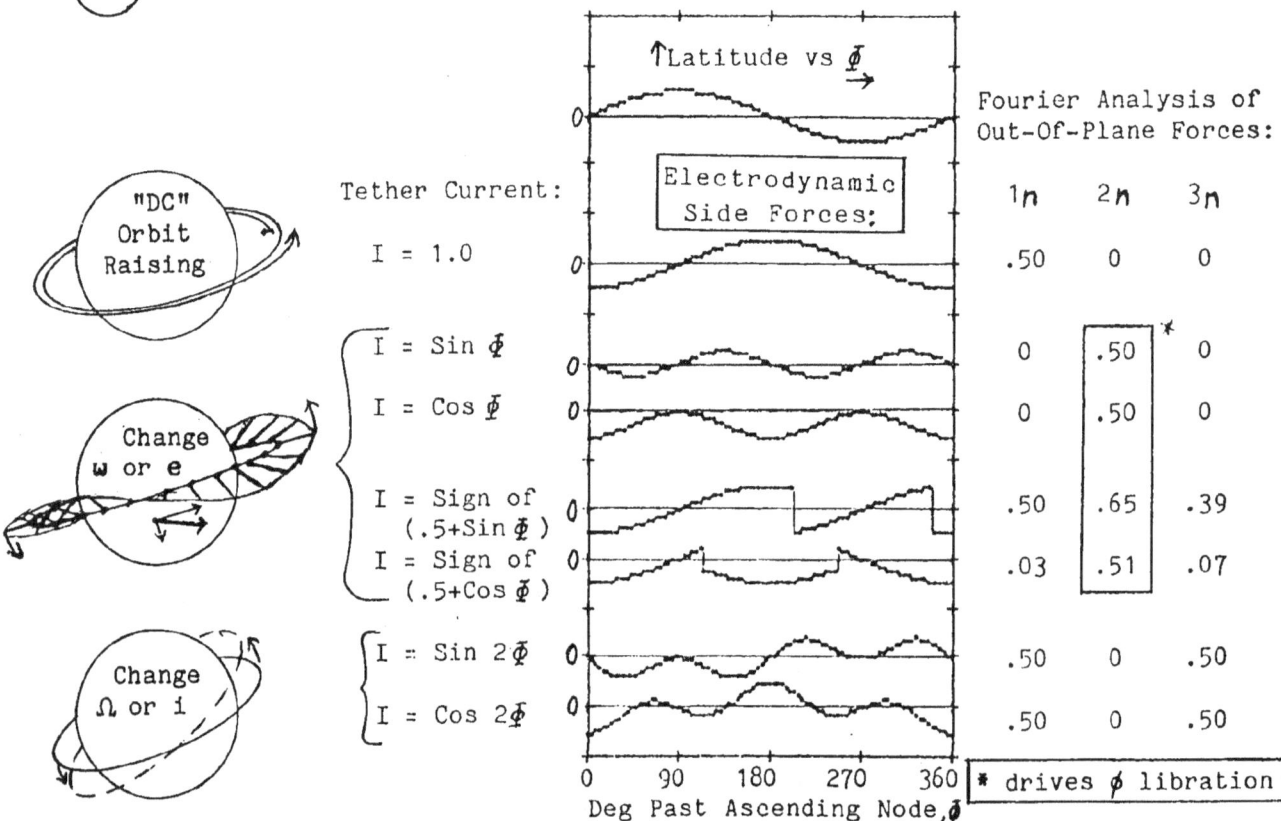

↑Latitude vs ϕ ⟶

Fourier Analysis of
Out-Of-Plane Forces:

"DC" Orbit Raising

Change ω or e

Change Ω or i

Tether Current:

Electrodynamic Side Forces:

	1n	2n	3n
I = 1.0	.50	0	0
I = Sin ϕ	0	.50	0
I = Cos ϕ	0	.50	0
I = Sign of (.5+Sin ϕ)	.50	.65	.39
I = Sign of (.5+Cos ϕ)	.03	.51	.07
I = Sin 2ϕ	.50	0	.50
I = Cos 2ϕ	.50	0	.50

0   90   180   270   360
Deg Past Ascending Node, ϕ

* drives ϕ libration

SECTION 6.0 SPACE SCIENCE AND TETHERS

6.1 Overview

Some scientific applications of tethers have been presented already in other sections of this handbook (see section 3 and 4). In this section we will illustrate the role that tethers can play in the future advancement of space science. We hope that this section will grow in the next editions.

According to the Non-advocate Tether Systems Applications Review (1993), chaired by Dr. M. Greenfield (see "contacts" Section), "...Space tether technology has the near-term potential to meet a broad range of science and technological aspects. The unique capabilities of tether technology enable the aquisition of science otherwise not achievable and can provide concepts for space applications...". Copies of report can be obtained either from the chairman or from the editors.

Space research with tethers has emphasized two particular applications: 1) Reaching otherwise unaccessible flight regions with downward deployed tethers; 2) Active experimentation with the surrounding plasma.

A good example of the effort carried on by the scientific community is the the workshop held in Ann Harbor, Michigan in July 1994. Copies of the Executive Summary can be obtained by Prof. B. Gilchrist (see "contacts" Section). The focus of this workshop was on how Ionospheric-Thermospheric-Mesospheric (ITM) Science can benefit from spaceborne tethers. NASA's sponsored TIMED mission promises to add substantially to the knowledge of the global response of the ITM region.

A multi -mass tether system could add many "in-situ" data on the effects of small scale spatial structures and its interactions (see "Applications" Section),. As the reports quotes "... Just as the advancement of remote sensing technology enabled the TIMED mission to be conceived, the ability of tethered payloads in space with spatial separations ranging for 1 Km to 100 km will enable a program of in-situ multiprobe diagnostics of the ITM region to be undertaken.". The workshop identified the following areas that would benefit from tethered spacecraft:

- Magnetospheric-Ionospheric coupling: Energy dissipation and configuration of three dimensional high latitude current systems.

- Effects of plasma structureson large and small scale electrodynamics.

- Ion-neutral momentum and energyexchange at different spatial scales.

- Momentum and energy transport processes by gravity waves.

- Thermospheric cooling (energy loss) through radiative emissions.

- The role of electromagnetic and electrostatic waves in energy transfer processes.

- The generation and flow of electrical currents in the ITM region

A task group chaired by prof. Heelis followed up the objectives laid out by the Michigan Workshop. The key science questions to be answered from a series of "in-situ" tether-aided observations in the lower thermosphere, highlighted significant advances as:

- Determination of the effective scales over which polarization electric fields are generated and how they map along the magnetic field lines.

- Determination of the wind effectivness in producing polarization fields and driving field-aligned currents.

- Identification of the type winds responsible for conductivity variations and those responsible for electric field generation.

- Assessment of gravity wave generators and of possible seed mechanism for F-region plasma instabilities.

- Assessment of the relecvance of thermospheric cooling to global change and impovment of prediction of the future physical characteristics in the thermosphere, mesosphere and stratosphere.

- Identification of the response of the lower ionosphere-thermosphere to large scale weather systems and transient phenomena associated with lightning.

The measurements that could address the above questions are listed in the following table.

Parameter	Dynamic Range	Accuracy	Resolution	Sample Interval
Neutral Atmospheric comp.	10^5-10^{11} cm^{-3}	$\leq \pm$ 10% and smaller for major species	$\Delta M/M=1$ at M=30 5%	<4 Km
Neutral Wind Vector	-500 to 500 m/s	$\pm$ 10%	1 m/s	<4 Km
Ion Composition	1 to 10^5 cm^{-3}	$\pm$ 10%	$\Delta M/M=1$ at M=16 1%	<4 Km Comp. <500 m Total
Ion Drift Velocity vector	-2 to +2 Km/s	$\pm$ 10%	1 m/s	<500 m
Ion/Electron/ Neutral Temp.	300 to 3000 K	$\pm$ 10%	50 K	<4 Km
Electric field Vector d.c.	-200 to +200 mV/m	$\pm$ 10%	0.05 mV/m	<4 Km
Current Density/ Magnetic field	-65 to +65 KnT	$\pm$ 0.1%	0.01%	<1 Km
FUV Imaging	10 R to 50 KR	0.5%	N/A	<1 Km
Energetic Particles	10 eV to 30 KeV 10^7 to 10^{10} cm^{-2} s^{-1} sr^{-1} eV^{-1}	$\pm$ 5%	N/A	<4 Km 30 deg pitch angle
IR Emissions 13-17.5 µm	2×10^{-9} to 5×10^{-8} W cm^{-2} sr^{-1}	10%	$\Delta R/R$ 3%	120 Km

4.17-6.25 μm	2×10^{-8} to 2×10^{-7}	5%		$\Delta R/R$ 0.4%	120 Km

More information on the instrumentation and the engineering aspects of this mission can be found in the section "Proposed Missions" (ATM Mission). A report entitled "Tether-based Investigation of the Ionosphere and Lower Thermosphere (TIILT)" has been prepared to present the scientific rationale behind this type of mission as well as the measurements and instrumentation. Copies of this report can be obtained by Prof. Heelis.

There are other missions, however, that would benefit from tethers . For example, AKTIVE spacecraft, launched by the former USSR in 1989, aimed at investigating VLF radiowave propagation and wave-particle interaction in the magnetosphere using a 10 KW VLF transmitter with a large loop antenna (20 m diameter). Electromagnetic effects occurring near the spacecraft were monitored by a coorbiting subsatellite, as shown in figure 6.1.

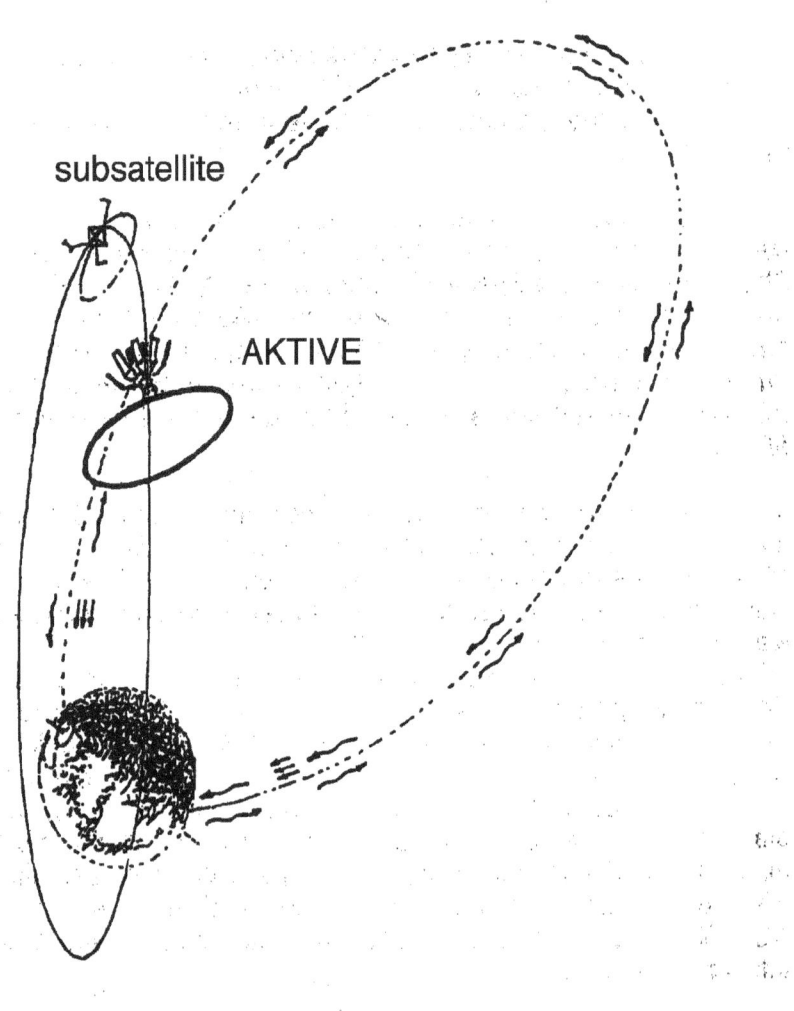

Figure 6.1 Aktive spacecraft and subsatellite

The primary objectives of the AKTIVE program were as follows:

1) Radiation Properties of the loop antenna.
2) Spatial structure of the electromagnetic fields in the near zone ($\leq$ 10 km).
3) Nonlinear effects in the near zone
4) Propagation of waves in the whistler mode, and their reflection from the ionosphere
5) Non-linear effects in whistler wave propagation
6) Precipitation of charged particles form the radiation belts due to interaction with VLF waves.
7) VLF emissions triggered from the orbiting AKTIVE transmitter.
8) Comparison with emissions triggered by ground based VLF transmitters.

Alas, AKTIVE encountered several technical problems and the program was terminated. Nevertheless, when the em-radiating properties of spaceborne tethers will be finally assessed, some of the above objectives , namely 4, 5 , 7 and 8 will greatly benefit. No further work has been done, however, in this direction. Some TSS investigations are currently addressing these questions.

6.2 Synergy

Some years ago, Lockheed-Martin, then Martin-Marietta, sponsored some studies to look into the synergy of tethers with other space missions, namely AFE (Aeroassist Flight Experiment), cancelled by NASA in 1991, and TIMED (Thermosphere-Ionosphere-Mesosphere Energetic Dynamics). Prof. Hurlbut (see "Contacts" Section) performed the study and the results are shown in tables 1 and 2, respectively.

AFE was a research "pathfinder" for a geosyncronous, lunar and planetary earth return aerobraking spacecraft. Prof. Hurlbut indicated that a tethered system could accomplish almost fifty percent of AFE objectives by exploring a much greater altitude range for a longer duration than AFE was supposed to fly.

The study on TIMED aimed at determinating which of its instruments could potentially fly on a pathfinder type tethered spacecraft. Note that the study of Lockheed-Martin on TIMED focused on one of its earliest configurations.
The major finding of this study was that a tethered spacecraft could possibly validate instruments which were operated in the 130-140 Km altitude range.

Table 1. AFE VS. Tethered System

AFE Flight Experiment	Tethered System Applicability
1. Forebody-Aerothermal Characterization Experiment (FACE)	Heat-flux and skin temperature measurements at all altitudes will provide thermal accommodation coefficients and validations of models/codes.
2. Radiative Heating Experiment (RHE)	Possibly applicable - Needs further study.
3. Wall Catalysis Experiment (WCE)	An extension of (1) to provide valuable catalytic vs. low catalytic gas/surface interaction data.
4. Base Flow Heating Experiment (BFHE)	Spherical afterbody data will differ from aerobrake geometry but will be very valuable with added Aerostabilizer instrument data.
5. Afterbody Radiometry Experiment (ARE)	Possibly applicable - Needs further study.
6. Alternate Thermal Protection Materials (ATPM)	Possibly applicable - Needs further study.
7. Heat Shield Performance (HSP)	Possibly applicable - Needs further study.
8. Pressure Distribution/Air Data System (PD/ADS)	Measurement of static/dynamic pressures at multiple satellite locations extremely valuable.
9. Aerodynamic Performance Experiment (APEX)	Satellite with Aerostabilizer will acquire extremely important aero characterization data over a wide altitude range.
10. Rarefield-Flow Aerodynamics Measurement Experiment (RAME) (RAME)	Measurements of momentum transfer characteristics and aero parameters (CD, CL, etc.) combined with (1) extremely valuable for validation of existing predictive analytical programs.
11. Plasma, Ion and Electron Concentration Experiment (PIECE)	Possibly applicable - Needs further study.
12. Microwave Reflectometer Ionization Sensor (MRIS)	Probably N/A
13. Aft Flow Ionization Sensor (MRIS)	Probably N/A

14. Ion Mass Spectrometer Experiment (IMSE)	Measurements of species and total density extremely important for atmospheric modeling.

Table 2. TIMED - Tethered Pathfinder Synergy

Timed Flight Experiment	Tethered System Pathfinder
1. Fabry-Perot Interferometer	Probably N/A - Requires more study.
2. Neutral Mass Spectrometer	Applicable for gas composition, temperatures and transverse winds.
3. Ion Mass Spectrometer	Applicability although ion composition and drift velocities of secondary importance.
4. Langmuir Probe	Applicable for measurement of electron temperatures and ion/electron densities.
5. Ion Drift Meter and Retarding Potentiometer	Applicable for measurement of ion temperatures, velocities and densities.
6. UV Spectrometer	Applicable for measurement of O3, NO temperatures, Noctilucent clouds, aerosols, and other minorconstituents.
7. Imaging Photometer	Possibly N/A - Requires more study.
8. Triaxial Accelerometer	Applicable as a high priority instrument.
9. Energetic Particle Analyzer	Probably N/A - Requires more study.
10. Global UV Airglow Imager	Probably N/A - Requires more study.
11. Solar EUV Spectrometer/UV Photometer	Probably N/A - Requires more study.
12. Vector Magnetometer	Applicable for magnetic field measurements
13. Near Infrared Spectrometer	Probably N/A - Requires more study.
14. Electric Field Detector/Plasma Wave Experiment	Probably N/A - Requires more study.
15. Infrared Limb Sounder	Probably N/A - Requires more study.
16. Fast Electron Spectrometer	Probably N/A - Requires more study.
17. Energetic Particle Spectrometer	Probably N/A - Requires more study.

SECTION 7.0 REFERENCES

7.1 General

Due to the large production of tether-related papers we have limited our search to works published in the scientific literature. We have also included the list of papers presented at the last tether conference held in Washington.

The proceedings of papers presented at the four international conferences on Tethers in Space as well as workshops can be found in:

- "Applications of Tethers in Space" Workshop held in Williamsburg, VA June 15-17, 1983. NASA Contract NAS8-35403.

- "Applications of Tethers in Space" Workshop held in Venice, Italy, October 15-17, 1985. NASA Conference Publication CP 2422.

- "International Conference of Tethers In Space", held in Arlington, VA, September 17-19, 1986. Proceedings published by the American Astronautical Society in *Advances in The Astronautical Sciences*, Vol. 62, 1987

- "Tether Dynamic Simulation Workshop", held in Arlington, VA Sept 16 1986. NASA Conference Publication CP 2458.

- "Space Tethers for Science in the Space Station Era", Conference held in Venice, October 4-8, 1987. Proceedings published by Societa' Italiana di Fisica, Bologna, Italy, 1988 (ISBN 88-7794-016-6).

- "Tethers In Space Toward Flight", Conference held in San Francisco, CA, May 17-19, 1989. Proceedings published by the American Institute of Aeronautics and Astronautics, 1989 (ISBN 0-930403-50-9).

- International Round Table on Tethers in Space", held in Noordwijk, The Netherlands, September 28-30, 1994. ESA WPP-081.

- "Fourth International Conference on Tethers in Space", held in Washington, DC, April 10-14, 1995. Published by Science and Technology Corporation, Hampton, VA.

7.2 Table of Contents of the Fourth International Conference on Tethers in Space

VOLUME I

217

220

VOLUME II

SMALL EXPENDABLE DEPLOYER SYSTEM (SEDS)

223

225

VOLUME III

DYNAMICS

FAR FUTURE

229

ed Elettronica-University of Genoa; Sergio Pagnan, Istituto di Automazione Navale - National Research Council of Italy; Luca Mina, Advanced Engineering Technology - Torre A Corte dei Lambruschini

Electrodynamic Interactions Between the PMG Tether and the Magneto-Ionic Medium of the Ionsphere   1891
        Mario D. Grossi and Robert D. Estes, Harvard-Smithsonian Center for Astrophysics; *James E. McCoy*, NASA, Johnson Space Center

Tether Current-Voltage Characteristics   1899
        R.C. Olsen, Chung-Jen Chang and Chia-Hwa-Chi, Naval Postgraduate School

APPENDIX A

Author Index   1923

APPENDIX B

Attendee List   1929

7.3 Bibliography

Alfven, H., "Spacecraft Propulsion: New Methods," Science, Vol. 176, p. 167-168, 14 Apr. 1972.

Allais, E. and Bergamaschi, S., "Dynamics of Tethered Satellites: Two Alternative Concepts for Retrieval," Meccanica, Vol. 14, No. 2, p. 103-111, June 1979.

Alpert, Ya. L., "On Some Electromagnetic Phenomena in the Tether Magnetoplasma Cloud," Nuovo Cimento C, Serie 1, Vol. 14C, September-October 1991.

Anderson, J., Wood, W., Siemers, P., Research at the Earth's Edge," Aerospace America, Vol. 26, No. 4, pp. 30-32, April 1988.

Anderson, K. S., Hagedon, Peter, "Control of Orbital Drift of Geostationary Tethered Satellites," Journal of Guidance, Control and Dynamics, Vol. 17, No. 1, p. 10-14, February 1994.

Anderson, L. A., "Tethered Elevator Design for Space Station," Journal of Spacecraft and Rockets, Vol. 29, p. 233-238, March-April 1992.

Anderson, W. W., "On Lateral Cable Oscillations of Cable-Connected Space Stations," NASA TN 5107, Langley Research Center, Hampton, Virginia, Mar. 1969.

Angrilli, F., Bianchini, G., Da Lio, M., Fanti, G.,"Modelling the mechanical Properties and Dynamics of the Tethers for the TSS-1 and TSS-2 Missions", ESA Journal, Vol. 12, p. 353-368, 1988

Arnold, D. A., "The Behavior of Long Tethers in Space," Journal of the Astronuatical Sciences, Vol. 35, No. 1, p. 3-18, January-March, 1987.

Ashenberg, J. and Lorenzini, E.C., "Dynamcs of a Dual-Probe Tethered System," Journal of Guidance, Control and Dynamics, Vol. 20, No. 6, p. 1265-1268, 1997.

Austin, F., "Nonlinear Dynamics of a Free-Rotating Flexibly Connected Double-Mass Space Station," Journal of Spacecraft and Rockets, Vol. 2, No. 6, p. 901-906, 1965.

Austin, F., "Torsional Dynamics of an Axially Symmetric, Two-Body Flexibly Connected Rotating Space Station," Journal of Spacecraft and Rockets, Vol. 2, No. 6, p. 626-628, 1965.

Bainum, P. M., and Evans, K. S., "Gravity-Gradient Effects on the Motion of Two Rotating Cable-Connected Bodies," AIAA Journal, Vol. 14, No. 1, p. 26-32, 1976.

Bainum, P. M., and Evans, K. S., "Three Dimensional Motion and Stability of Two Rotating Cable-Connected Bodies," Journal of Spacecraft and Rockets, Vol. 12, No. 4, p. 242-250, 1975.

Bainum, P. M., and Kumar, V. K., "Optimal Control of the Shuttle-Tethered System," Acta Astronautica, Vol. 7, No. 12, p. 1333-1348, May 1980.

Bainum, P. M., Diarra, C. M., and Kumar, V. K., "Shuttle-Tethered Subsatellite System Stability with a Flexible Massive Tether," AIAA J. Guidance, Control and Dynamics, Vol. 8, No. 2, p. 230-234, Mar.-Apr. 1985.

Bainum, P. M., Harkness, R. E. and Stuiver, W., "Attitude Stability and Damping of a Tethered Orbiting Interferometer Satellite System," The Journal of Astronautical Sciences, Vol. 19, No. 5, p. 364-389, Mar-Apr. 1972.

Baker, W. P., Dunkin, J. A., Galaboff, Z. J., Johnson, K. D., Kissel, R. R., Rheinfurth, M. H., and Siebel, M.P.L., "Tethered Subsatellite Study," NASA TM X-73314, NASA/MSFC, Mar. 1976.

Banerjee, A. K., and Kane, T. R., "Tether Deployment Dynamics," The Journal of Astronautical Sciences, Vol. 30, No. 4, p. 347-365, Oct-Dec 1982.

Banerjee, A. K., and Kane, T. R., "Tethered Satellite Retrieval with Thruster Augmented Control" Journal of Guidance, Control and Dynamics, Vol. 7, No. 1, p. 45-50, 1984.

Banerjee, A. K., "Dynamics of Tethered Payloads with Deployment Rate Control," Journal of Guidance, Control and Dynamics, Vol. 13, p. 759-762, July-August 1990.

Banks, P., et al, "The Tethered Satellite System; Final Report from the Facility Requirements Definition Team," NAS8-33383, MSFC, May 1980.

Bekey, I. "Tethers Open New Space Options," Astronautics and Aeronautics, Vol. 21, No. 4, p. 33-40, Apr. 1983.

Bekey, I., and Penzo, P. A., "Tether Propulsion," Aerospace America, Vol. 24, No. 7, p. 40-43, July 1986.

Beletskii, V. V. and Levin, E. M., "Dynamics of Space Tether Systems," Advances in the Astronautical Sciences, Vol. 83.

Beletskii, V. V., and Guivertz, M., "The Motion of an Oscillating Rod Subjected to a Gravitational Field," Kosmitcheskie Issledovania, Vol. 5, No.6, 1967.

Beletskii, V. V., and Levin, E. M., "Dynamics of the Orbital Cable System," Acta Astronautica, Vol. 12, No. 5, p. 285-291, 1985.

Beletskii, V. V., and Levin, E. M., "Stability of a Ring of Connected Satellites," Acta Astronautica, Vol. 12, No. 10, p. 765-769, 1985.

Beletskii, V. V., and Navikova, E. T., "On the Relative Motion of Two Cable-Connected Bodies in Orbit," Cosmic Research. Vol. 7, No. 3, p. 377-384, 1969.

Beletskii, V. V., "On the Relative Motion of Two Cable-Connected Bodies in Orbit-II," Cosmic Research, Vol. 7, No. 6, p. 827-840, 1969.

Bergamaschi, S., "Tether Motion after Failure," Journal of Astronautical Sciences, Vol. 30, No. 1, p. 49-59, Jan.-Mar. 1982.

Bergamaschi, S. and Catinaccio, A., "Further Developments in the Harmonic Analysis of TSS-1," Journal of the Astronautical Sciences, Vol. 40, No. 2, p. 189-201, April-June 1992.

Bergamaschi, S. and Bonon, F., "Coupling of Tether Lateral Vibration and Subsatellite Attitude Motion," Journal of Guidance, Control and Dynamics, Vol. 15, No. 5, p. 1284-1286, September-October 1992.

Bergamaschi, S. and Bonon, F., "Equilibrium Configurations in a Tethered Atmospheric Mission," Acta Astronautica, Vol. 29, No. 5, p. 333-339, May 1993.

Bergamaschi, S., Bonon, F. and Legnami, M., "Spectral Analysis of Tethered Satellite System-Mission 1 Vibrations," Journal of Guidance, Control and Dynamics, Vol. 18, No. 3, p. 618-624, June 1995.

Bergamaschi, S., Zanetti, P. and Zottarel, C., "Nonlinear Vibrations in the Tethered Satellite System-Mission 1," Journal of Guidance, Control and Dynamics, Vol. 19, No. 2, p. 289-296, April 1996.

Birch, P., "Orbital Ring Systems and Jacob's Ladders," Journal of British Interplanetary Society, Vol. 35, No. 11, p. 474-497, Nov. 1982 (Part 1), Vol. 36, No. 3, p. 115-128, Dec. 1983 (Part 2).

Blinov, A. P., Dosybekov, K., "Perturbed Motions of a Dumbbell in a Central Newtonian Force Field," ISSN 0010-9525, Cosmic Research, Jan. 1988.

Bolotina, N. E., and Vilke, V. G., "Stability of the Equilibrium Positions of a Flexible, Heavy Fiber Attached to a Satellite in a Circular Orbit," Cosmic Research, Vol. 16, No. 4, p. 506-510, Jan. 1979.

Breakwell, J. V., Gearhart, J. W., "Pumping a Tethered Configuration to Boost its Orbit Around an Oblate Planet," Journal of the Astronuatical Sciences, Vol. 35, No. 1, p. 19-40, January-March, 1987.

Breakwell, J. V., "Stability of an Orbiting Ring," Journal of Guidance and Control, Vol. 4, No. 2, p. 197-200, 1981.

Brown, K. G., Melfi, L. T., Jr., Upchurch, B.T. and Wood, G. M. Jr., "Downward-Deployed Tethered Satellite Systems, Measurement Techniques, and Instrumentation - A Review," Journal of Spacecraft and Rockets, Vol. 29, No. 5, p. 671-677, September-October 1992.

Bschorr, O., "Controlling Short-Tethered Satellites," Acta Astronautica, Vol. 7, p. 567-573, May 1980.

Carroll, J. A., "Tether Applications in Space Transportation," Acta Astronautica, Vol. 13, No. 4, p. 165-174, 1986.

Chang, C.L., Satya-Narayana, P., Drobot, A.T., Papadopoulos, L. and Lipatov, A.S., "Hybrid Simulations of Whistler Waves Generation and Current Closure by a Pulsed Tether in the Ionosphere," Geophysical Research Letters, Vol. 21, No. 11, p. 1015-1018, June 1, 1994.

Childs, D. W., and Hardison, T. L., "A Movable-Mass Attitude Stabilization System for Cable-Connected Artificial-g Space Stations," Journal of Spacecraft and Rockets, Vol. 11 No. 3, p. 165-172, 1974.

Chobotov, V. A., "A Synchronous Satellite at Less Than Synchronous Altitude," Journal of Spacecraft and Rockets, Vol. 13, No. 2, p. 126-128, 1976.

Chobotov, V. A., "Gravity-gradient Excitation of a Rotating Cable-Counterweight Space Station in Orbit," Journal of Applied Mechanics, Vol. 30, No. 4, p. 547-554, Dec. 1963.

Chu, C., and Gross, R., "Alfven Waves and Induction Drag on Long Cylindrical Satellites," AIAA Journal, Vol. 4, p. 2209, 1966.

Collar, A., and Flower, J., "A (Relatively) Low Altitude 24-Hour Satellite," J. British Interplanetary Society, Vol. 22, p. 442-457, 1969.

Colombo, G., Gaposchkin, E. M., Grossi, M. D., and Weiffenbach, G. C., "The 'Skyhook': A Shuttle-Borne Tool for Low Orbital Altitude Research," Meccanica, Vol. 10, No. 1, Mar. 1975.

Corso, G. J., "A Proposal to Use an Upper Atmosphere Satellite Tethered to the Space Shuttle for the Collection of Micro-Meteoric Material," J. of the British Interplanetary Society, Vol. 36, p. 403-408, 1983.

Cotellessa, A. and DeMatteis, G., "Passive Stabilization of a Tethered System in Low Earth Orbit," Acta Astronautica, Vol. 29, No. 3, p. 169-180, March 1993.

Crist, S. A., and Eisley, J. G., "Cable Motion of a Spinning Spring-Mass System in Orbit," Journal of Spacecraft and Rockets, Vol. 7, No. 11, p. 1352-1357, 1970.

Davis, W.R. and Banerjee, A.K., "Libration Damping of a Tethered Satellite by Yo-Yo Control with Angle Measurement," Journal of Guidance, Control and Dynamics, Vol. 13, p. 370-374, March-April 1990.

Decou, A.B., "Orbital Dynamics of the Hanging Tether Interferometer," Journal of Guidance, Control and Dynamics, Vol. 14, p. 1309-1311, November-December 1991.

DeMatteis, G. and DeSocio, Luciano M., "Stability of a Tethered Satellite Subjected to Stochastic Forces," Acta Astronautica, Vol. 25, p. 61-66, February 1991.

DeMatteis, G., "Dynamics of a Tethered Satellite in Elliptical, Non-Equatorial Orbits," Journal of Guidance, Control and Dynamics, Vol. 15, No. 3, p. 621-626, May-June 1992.

Denig, W.F., Maynard, N.C., Burke, W.J. and Maehlum, B.N., "Electric Field Measurements During Supercharging Events on the MAIMIK Rocket Experiment," Journal of Geophysical Research, Vol. 96, p. 3601-3610, March 1, 1991.

Dobrowolny, M., Arnold, D., Colombo, G., and Grossi, M., "Mechanisms of Electrodynamic Interactions with a Tethered Satellite System and the Ionosphere," Reports in Radio and Geoastronomy, No. 6, Aug., 1979.

Dobrowolny, M., "Electrodynamics of Long Metallic Tethers in the Ionospheric Plasma," Radio Science, Vol. 13, p. 417, 1978.

Dobrowolny, M. and Melchioni, E., "Electrodynamic Aspects of the First Tethered Satellite Mission," Journal of Geophysical Research, Vol. 98, No. A8, p. 13,761-13,778, August 1, 1993.

Dobrowolny, M. (ed.), "Special TSS-1 Issue," Il Nuovo Cimento, Vol. 17c, January-February, 1994.

Donohue, D.J., Neubert, T. and Banks, P.M., "Estimating Radiated Power from a Conducting Tethered Satellite System," Journal of Geophysical Research, Vol. 96, p. 21,245-21,253, December 1, 1991.

Drell, S. D., Foley, H. M., and Ruderman, M. A., "Drag and Propulsion of Large Satellites in the Ionosphere: An Alfven Propulsion Engine in Space," Journal of Geophysical Research, Vol. 70, No. 13, p. 3131-3145, July 1965.

Ebner, S. G., "Deployment Dynamics of Rotating Cable-Connected Space Stations," Journal of Spacecraft and Rockets, Vol. 7, No. 10, p. 1274-1275, 1970.

Estes, R. D., "Alfvén Waves from an Electrodynamic Tethered Satellite System," Journal of Geophysical Research, Vol. 93, A2, p. 945, 1988.

Fleurisson, E J., VonFlotow, Andreas H. and Pines, Darryll J., "Trajectory Design, and Feedback Stabilization of Tethered Spacecraft Retrieval," Journal of Guidance, Control and Dynamics, Vol. 16, No. 1, p. 160-167, January-February 1993.

Fujii, H., Uchiyama, K. and Kokubun, K., "Mission Function Control of Tethered Subsatellite Deployment/Retrieval - In-Plane and Out-of-Plane Motion," Journal of Guidance, Control and Dynamics, Vol. 14, p. 471-473, March-April 1991.

Fujii, H., Kokubun, K., Uchiyama, K. and Suganuma, T., "Deployment/Retrieval Control of a Tethered Subsatellite Under Aerodynamic Effect of Atmosphere," Journal of the Astronautical Sciences, Vol. 40, No. 2, p. 171-188, April-June 1992.

Fujii, H. A. and Anazawa, S., "Deployment/Retrieval Control of Tethered Subsatellite Through an Optimal Path," Journal of Guidance, Control and Dynamics, Vol. 17, No. 6, p. 1292-1298, December 1994.

Furta, S.D., "On the Instability of 'Folded' Equilibria of a Flexible Nonstretchable Thread Attached to the Satellite in a Circular Orbit," Celestial Mechanics and Dynamical Astronomy, Vol. 53, No. 3, p. 255-266, 1992.

Gioulekas, A. and Hastings, D.E., "Role of Current Driven Instabilities in the Operation of Plasma Contactors Used with Electrodynamic Tethers," Journal of Propulsion and Power, Vol. 6, p. 559-566, September-October 1990.

Glickman, R. E. and Rybak, S. C., "Gravity Gradient Enhancement During Tethered Payload Retrieval," Journal of the Astronuatical Sciences, Vol. 35, No. 1, p. 57-74, January-March, 1987.

Grassi, M. and Cosmo, M. L., "Attitude Dynamics of the Small Expendable-Tether Deployment System," Acta Astronautica, Vol. 36, No. 3, p. 141-148, August 1995.

Grassi, M. and Cosmo, M. L., "Atmospheric Research with the Small Expendable Deployer System: Preliminary Analysis," Journal of Spacecraft and Rockets, Vol. 33, No. 1, p. 70-78, January-February 1996.

Greene, M., Rupp, C. C., Walls, J., Wheelock, D. and Lorenzoni, A., "Feasibility Assessment of the Get-Away Tether Experiment," Journal of the Astronuatical Sciences, Vol. 35, No. 1, p. 75-96, January-March, 1987.

Greene, M. E. and Denney, T. S., "On State Estimation for an Orbiting Single Tether System," IEEE Transactions on Aerospace and Electronic Systems, Vol. 27, p. 689-695, July 1991.

Greene, M. E. and Denney, T. S., Jr., "Real-Time Estimator for Control of an Orbiting Single Tether System," IEEE Transactions on Aerospace and Electronic Systems, Vol. 27, p. 880-883, November 1991.

Greene, M.E., Carter, J.T. and Walls, J.L., "Linear Adaptive Control of a Single-Tether System," International Journal of Adaptive Control and Signal Processing, Vol. 6, No. 1, p. 1-17, January 1992.

Grossi, M. D., "A ULF Dipole Antenna on a Spaceborne Platform of the PPEPL Class," Report for NASA contract NAS8-28203, May, 1973.

Grossi, M. D., "Spaceborne Long Vertical Wire as a Self-Powered ULF/ELF Radiator," IEEE Journal of Oceanic Engineering, Vol. OE-9, No. 3, p. 211-213, July 1984.

Gullahorn, G., Fuligni, F., Grossi, M. D., "Gravity Gradiometry from the Tethered Satellite System," IEEE Transportation Geoscience Remote Sensing, Vol. GE 23, p. 531-540, 1985.

Gwaltney, D. A. and Greene, M. E., "Ground-Based Implementation and Verification of Control Laws for Tethered Satellites," Journal of Guidance, Control and Dynamics, Vol. 15, p. 271-273, January-February 1992.

He, X. and Powell, J. D., "Tether Damping in Space," Journal of Guidance, Control and Dynamics, Vol. 13, p. 104-112, January-February 1990.

Humble, R. W., "Two Dimensional Tethered Satellite Attitude Dynamics," Journal of the Astronautical Sciences, Vol. 38, p. 21-27, January-March 1990.

Hurlbut, F.C. and Potter, J.L., "Tethered Aerothermodynamic Research Needs," Journal of Spacecraft and Rockets, Vol. 28, p. 50-57, January-February 1991.

Ionasescu, R., and Penzo, P. A., "Space Tethers," British Interplanetary Society, Spaceflight, Vol. 30, No. 5, May 1988.

Isaacs, J. D., Vine, A. C., Bradner, H., and Bachus, G. E., "Satellite Elongation into a True 'Sky-Hook'," Science, Vol. 151, p. 682, 683 (Feb. 1966), Vol. 152, p. 800, Vol. 158, p. 946, 947, Nov. 1967.

James, H.G., "Wave Results from OEDIPUS A (Rocket Sounding of Plasma Dynamics in Auroral Ionosphere),"

Advances in Space Research, Vol. 13, No. 10, p. 5-13, October 1993.

Kalaghan, P., Arnold, D. A. , Colombo, G., Grossi, M., Kirschner, L. R., and Orringer, O., "Study of the Dynamics of a Tethered Satellite System (Skyhook)," NASA Contract NAS8-32199, SAO Final Report, Mar. 1978.

Kane, T. R., "A New Method for the Retrieval of the Shuttle-Based Tethered Satellite," Journal of the Astronautical Sciences, Vol. 32, No. 3, p. 351-354, July-Sept. 1984.

Kane, T. R., and Levinson, D. A., "Deployment of a Cable-Supported Payload from an Orbiting Spacecraft," Journal of Spacecraft and Rockets, Vol. 14, No. 7, p. 409-413, 1977.

Katz, I., Lilley, J.R. Jr., Greb, A., McCoy, J.E., Galofaro, J. and Ferguson, D.C., "Plasma Turbulence Enhanced Current Collection - Results from the Plasma Motor Generator Electrodynamic Tether Flight," Journal of Geophysical Research, Vol. 100, No. A2, February 1995.

Kelly, W. D., "Delivery and Disposal of a Space Shuttle External Tank to Low-Earth Orbit," Journal of the Astronautical Sciences, Vol. 32, No. 3, p. 343-350, July-Sept. 1984.

Kerr, W. C., and Abel, J. M., "Traverse Vibrations of a Rotational Counterweighted Cable of Small Flexural Rigidity," AIAA Journal, Vol. 9, No. 12, p. 2326-2332, 1971.

Keshmiri, M., Misra, A.K. and Modi, V.J., "General Formulation for N-Body Tethered Satellite System Dynamics," Journal of Guidance, Control and Dynamics, Vol. 19, No. 1, p. 75-83, February 1996.

Kim, E. and Vadali, S. R., "Modeling Issues Related to Retrieval of Flexible Tethered Satellite System," Journal of Guidance, Control and Dynamics, Vol. 18, No. 5, p. 1169-1176, October 1995.

Kline-Schoder, R. J. and Powell, J.D., "Precision Attitude Control for Tethered Satellites," Journal of Guidance, Control and Dynamics, Vol. 16, No. 1, p. 168-174, January-February 1993.

Kumar, K., Kumar, R. and Misra, A.K., "Effects of Deployment Rates and Librations on Tethered Payload

Raising," Journal of Guidance, Control and Dynamics, Vol. 15, No. 5, p. 1230-1235, September-October 1992.

Kumar, K., "Geosynchronous Satellites at Sub-Synchronous Altitudes," Acta Astronautica, Vol. 29, No. 3, p. 149-151, March 1993.

Kyroudis, G. A., and Conway, B. A., "Advantages of Using an Elliptically-Orbiting Tethered-Dumbbell System for a Satellite Transfer to Geosynchronous Orbit," submitted to Journal of Guidance, Control and Dynamics, 1987.

Lemke, E. H., "On a Lunar Space Elevator," Acta Astronautica, Vol. 12, No. 6, 1985.

Lemke, L. G., Powell, J. D., He, X., "Attitude Control of Tethered Spacecraft," Journal of the Astronuatical Sciences, Vol. 35, No. 1, p. 41-56, January-March, 1987.

Levin, E. M., "Stability of the Stationary Motions of an Electromagnetic Tether System in Orbit," ISSN 0010-9525, Cosmic Research, Jan. 1988.

Levin, E. M., "Stability of the Time-Independent Tethered Motions of Two Bodies in Orbit Under the Action of Gravitational and Aerodynamic Forces," Translated from Komicheskie Issledovaniya, Vol. 22, No. 5, p. 675-682, Sept.-Oct. 1984.

Levin, E.M., "Nonlinear Oscillations of Space Tethers," Acta Astronautica, Vol. 32, No. 5, p. 405-408, May 1994.

Levin, E.M., "Nearly-Uniform Deployment Strategy for Space Tether Systems," Acta Astronautica, Vol. 32, No. 5, p. 399-403, May 1994.

Li, Z. and Bainum, P. M., "On the Development of Control Laws for an Orbiting Tethered Antenna/Reflector System Test Scale Model," Journal of Intelligent Material Systems and Structures, Vol. 4, No. 3, p. 343-353, July 1993.

Liaw, D. and Abed, E., H., "Stabilization of Tethered Satellites During Station Keeping," IEEE Transactions on Automatic Control, Vol. 35, p. 1186-1196, November 1990.

Lips, K. W, and Modi, V. J., "General Dynamics of a Large Class of Flexible Satellite Systems," Acta Astronautica, Vol. 7, p. 1349-1360, 1980.

Lips, K. W., Modi, V. J., "Transient Attitude Dynamics of Satellites with Deploying Flexible Appendages," Acta Astronautica, Vol. 5, p. 797-815, Oct. 1978.

Longuski, J.M., Puig-Suari, J., and Mechalas, J., "Aerobraking Tethers for the Exploration of the Solar System," Acta Astronautica, Vol. 35, No. 2/3, pp. 205-214, 1995.

Longuski, J.M., Puig-Suari, J., Tsiotras, P., and Tragesser, S., "Optimal Mass for Aerobraking Tethers," Acta Astronautica, Vol. 35, No. 8, pp. 489-500, 1995.

Lorenzini, E. C., "A Three-Mass Tethered System for Micro-g/Variable-g Applications," Journal of Guidance, Control, and Dynamics, Vol. 10, No. 3, May-June 1987. Also in the Russian Journal of Aeronautics/Space Technology, No. 12, Dec. 1987.

Lorenzini, E. C., Cosmo, M., Vertrella, S., and Moccia, A., "Dynamics and Control of the Tether Elevator/Crawler System," to appear in the Journal of Guidance, Control and Dynamics, 1988.

Lorenzini, E.C., MD, Grossi, and M. Cosmo, " Low Altitude Tethered Mars Probe," Acta Astronautica, Vol 21, No.1, 1990, pp. 1-12.

Lorenzini, E. C., Gullahorn, G. E., and Fuligni, F., "Recent Developments in Gravity Gradiometry from the Space-Shuttle-Borne Tethered Satellite System," Journal of Applied Physics, Vol. 63, No. 1, p. 216-223, Jan. 1988.

Lorenzini, E. C., "Novel Tether-Connected Two-Dimensional Structures for Low Earth Orbits," Journal of Astronautical Sciences, Vol. 36, No. 4, Oct.-Dec. 1988.

Lorenzini, E.C., Sullivan, J. D. and Post, R. S., "New Techniques for Collecting Data Around the Space Station," Journal of the Astronautical Sciences, Vol. 38, p. 121-141, April-June 1990.

Lorenzini, E.C., Bortolami, S.B., Rupp, C.C., Angrilli, F., "Control and Flight Performance of Tethered Satellite Small Expendable Deployment System-II," Journal of Guidance, Control, and Dynamics, Vol. 19, No.5, p. 1148-1156, September-OCtober 1996.

Lüttgen, A. and Neubauer, F.M., "Generation of Plasma Waves by a Tethered Satellite Elongated in the Direction of Flight for Arbitrary Oblique Geometry," Journal of Geophysical Research, Vol. 99, A12, p. 23,349, 1994.

Maiorov, V. A., Popov, V. I., Yanov, I. O., "Analysis of the Dynamics of a System Using Gravitational-Gradient and Gyroscopic Principles of Stabilization," Cosmic Research., Vol. 18, No. 4, p. 348-356, Jul.-Aug. 1980.

Manaziruddin, and Singh, R.B., "Effects of Small External Forces on the Planar Oscillation of a Cable Connected Satellites System," Celestial Mechanics and Dynamical Astronomy, Vol. 53, No. 3, p. 219-226, 1992.

Martinez-Sanchez, M., Hastings, D. E., "A Systems Study of a 100 KW Electrodynamic Tether," Journal of the Astronuatical Sciences, Vol. 35, No. 1, p. 75-96, January-March, 1987.

Martinez-Sanchez, M., and Gavit, S. , "Orbit Modifications Using Forced Tether Length Variations." Journal of Guidance, Control and Dynamics, Vol. 10, No. 3, p. 233-241, 1987.

McComas, D. J., Spence, H. E., Karl, R. R., Horak, H. G., Wilkerson, T. D., "Bistatic LIDAR Experiment Proposed for the Shuttle/Tethered Satellite System Missions," Review of Scientific Instruments, Vol. 56, No. 5, p. 670-673, May 1985.

Misra, A. K., and Modi, V. J., "Deployment and Retrieval of Shuttle Supported Tethered Satellites," Journal of Guidance, Control and Dynamics, Vol. 5, No. 3, p. 278-285, 1982.

Misra, A. K., Modi, V. J., "The Influence of Satellite Flexibility on Orbital Motion," Celestial Mechanics, Vol. 17, p. 145-165, Feb. 1978.

Moccia, A. and Vetrella, S., "A Tethered Interferometric Synthetic Aperture Radar (SAR) for a Topographic Mission," IEEE Transactions on Geoscience and Remote Sensing, Vol. 30, p. 103-109, January 1992.

Moccia, A., Vetrella, S. and Grassi, M., "Attitude Dynamics and Control of a Vertical Interferometric Radar Tethered Altimeter," Journal of Guidance, Control and Dynamics, Vol. 16, No. 2, p. 264-269, March-April 1993.

Modi, V. J. , Chang-fu, G., and Misra, A. K., "Effects of Damping on the Control Dynamics of the Space Shuttle Based Tether System," Journal of the Astronautical Sciences, Vol. 31, No. 1, p. 135-149, Jan.-Mar. 1983.

Modi, V. J., and Misra, A. K., "On the Deployment Dynamics of Tether Connected Two-Body Systems," Acta Astronautica, Vol. 6, No. 9, p. 1183-1197, 1979.

Modi, V. J., and Misra, A. K., "Orbital Perturbations of Tethered Satellite Systems," Journal of the Astronautical Sciences, Vol. 25, No. 3, p. 271-278, July-Sept. 1977.

Modi, V.J., Chang-Fu, G., Misra, A. K., and Xu, D. M., "On the Control of the Space Shuttle Based Tether System," Acta Astronautica, Vol. 9. No. 6-7, p. 437-443, 1982.

Modi, V.J., Lakshmanan, P.K. and Misra, A.K., "On the Control of Tethered Satellite Systems," Acta Astronautica, Vol. 26, No. 6, p. 411-423, June 1992.

Modi, V.J., Bachmann, S. and Misra, A.K., "Dynamics and Control of a Space Station Based Tethered Elevator System," Acta Astronautica, Vol. 29, No. 6, p. 429-449, June 1993.

Moravec, H., "A Non-Synchronous Orbital Skyhook," Journal of the Astronautical Sciences, Vol. 25, No. 4, p. 307-322, Oct.-Dec. 1977.

Myers, N.B., Ernstmeyer, J., McGill, P., Fraser-Smith, A.C., Raitt, W.J. and Thompson, D.C., "Ground-Based VLF Measurements During Pulsed Electron Beam Emissions in the Ionosphere," Advances in Space Research, Vol. 13, No. 10, p. 99-102, October 1993.

Netzer, E. and Kane, T. R., "An Alternate Approach to Space Missions Involving a Long Tether," Journal of the Astronautical Sciences, Vol. 40, No. 3, p. 313-327, July-September 1992.

Netzer, E. and Kane, T. R., "Deployment and Retrieval Optimization of a Tethered Satellite System," Journal of Guidance, Control and Dynamics, Vol. 16, No. 6, p. 1085-1091, November-December 1993.

Netzer, E. and Kane, T., "Estimation and Control of Tethered Satellite Systems," Journal of Guidance, Control and Dynamics, Vol. 18, No. 4, p. 851-858, August 1995.

Neubert T., Gilchrist, B. and Ungstrup, E., "AMPAS - A New Active Experiment Mission (Active Magnetospheric Particle Acceleration Satellite," Advances in Space Research, Vol. 15, No. 12, p. 3-12, June 1995.

Nixon, D. D., "Dynamics of a Spinning Space Station with a Counterweight Connected by Multiple Cables," Journal of Spacecraft and Rockets, Vol. 9, No. 12, p. 896-902, 1972.

Oberhardt, M.R., Hardy, D.A., Thompson, D.C., Raitt, W.J., Melchioni, E., Bonifazi, C. and Gough, M.P., "Positive Spacecraft Charging as Measured by the Shuttle Potential and Return Electron Experiment," IEEE Transactions on Nuclear Science, Vol. 40, No. 6, p. 1532-1541, December 1993.

Pasca M. and E. C. Lorenzini, "Optimization of a Low Altitude Tethered Probe for Martian Atmospheric Collection", The Journal of the Astronautical Sciences, Vol. 44, No.2, 1996, pp.191-205

Pasca, M. Pignataro, M. and Luongo, A., "Three-Dimensional Vibrations of Tethered Satellite System," Journal of Guidance, Control and Dynamics, Vol. 14, p. 312-320, March-April 1991.

Pearson, J., "Anchored Lunar Satellites for Cislunar Transportation and Communication," Journal of the Astronautical Sciences, Vol. 27, No. 1, Jan.-Mar. 1979.

Pearson, J., "The Orbital Tower: A Spacecraft Launcher Using the Earth's Rotational Energy," Acta Astronautica, Vol. 2, No. 9/10, p. 785-799, Sept.-Oct . 1975.

Pelaez, J., "On the Dynamics of the Deployment of a Tether from an Orbiter. I - Basic Equations," Acta Astronautica, Vol. 36, No. 2, p. 113-122, 1995.

Pelaez, J., "On the Dynamics of the Deployment of a Tether from an Orbiter. II - Exponential Deployment," Acta Astronautica, Vol. 36, No. 6, p. 313-335, 1995.

Pengelley, C. D., "Preliminary Survey of Dynamic Stability of Cable-Connected Spinning Space Station," Journal of Spacecraft and Rockets, Vol. 3, No. 10, p. 1456-1462, Oct. 1966.

Penzo, P. A., and Mayer, H. L., "Tethers and Asteroids for Artificial Gravity Assist in the Solar System," Journal of Spacecraft and Rockets, Vol. 23, No. 1, Jan.-Feb. 1986.

Pines, D.J., VonFlotow, A.H. and Redding, D.C., "Two Nonlinear Control Approaches for Retrieval of a Thrusting Tethered Subsatellite," Journal of Guidance, Control and Dynamics, Vol. 13, p. 651-658, July-August 1990.

Polites, M.E., "Reconstructing Tethered Satellite Skiprope Motion by Bandpass Filtering Magnetometer Measurements," Journal of Dynamic Systems, Measurement and Control, Vol. 114, No. 3, p. 481-485, September 1992.

Pradhan, S., Modi, V. J. and Misra, A. K., "On the Inverse Control of the Tethered Satellite System," Journal of the Astronautical Sciences, Vol. 43, No. 2, p. 179-193, June 1995.

Pringle, R., "Exploration of Nonlinear Resonance in Damping an Elastic Dumb-Bell Satellite," AIAA Journal, No. 7, p. 1217-1222, 1968.

Puig-Suari, J., Logunski, J. M. and Tragesser, S. G., "Aerocapture with a Flexible Tether," Journal of Guidance, Control and Dynamics, Vol. 18, No. 6, p. 1305-1312, December 1995.

Puig-Suari, J., Logunski, J. M. and Tragesser, S. G., "A Tether Sling for Lunar and Interplanetary Exploration," Acta Astronautica, Vol. 36, No. 6, p. 291-295, September 1995.

Quadrelli, B.M. and Lorenzini, E.C., "Dynamics and Stability of a Tethered Centrifuge in Low Earth Orbit," Journal of the Astronautical Sciences, Vol. 40, p. 3-25, January-March 1992.

Raitt, W.J., Ernstmeyer, James, Myers, Neil B., White, A.B., Sasaki, Susumu, Oyama, Koh-ichiro, Kawashima, Nobuki, Fraser-Smith, Anthony C., Gilchrist, Brian E. and Hallinan, Thomas J., "VLF Wave Experiments in Space Using a Modulated Electron Beam," Journal of Spacecraft and Rockets, Vol. 32, No. 4, p. 670-679, August 1995.

Ross, J., "Super-strength Fiber Applications," Astronautics and Aeronautics, Dec. 1977.

Rupp, C. C., and Laue, J. H., "Shuttle/Tethered Satellite System," The Journal of the Astronautical Sciences, Vol. 26, No. 1, p. 1-17, Jan. 1978.

Rupp, C. C., et al, "Shuttle/Tethered Satellite System Conceptual Design Study," NASA TMX-73365, MSFC, Alabama, Dec. 1976.

Samanta R., R.I. and Hastings, D.E., "Theory of Plasma Contactor Neutral Gas Emissions for Electrodynamic Tethers," Journal of Spacecraft and Rockets, Vol. 29, No. 3, p. 405-414, May-June 1992.

Sanmartin, J.R. and Lam, S.H., "Far-Wake Structure in Rarefield Plasma Flows Past Charged Bodies," Phys. Fluids, 14, p. 62, 1971.

Sanmartin, J.R., Martinez-Sanchez, M. and Ahedo, E., "Bare Wire Anodes for Electrodynamic Tethers," Journal of Propulsion and Power, Vol. 9, No. 3, p. 353-360, May-June 1993.

Sanmartin, J.R. and Martinez-Sanchez, M., "The Radiation Impedance of Orbiting Conductors," Journal of Geophysical Research, Vol. 100, No. A2, p. 1677-1686, February 1, 1995.

Santangelo, A. D. and Johnson, G. E., "Optimal Wing Configuration of a Tethered Satellite System in Free Molecular Flow," Journal of Spacecraft and Rockets, Vol. 29, No. 5, p. 668-670, September-October 1992.

Singh, N. and Vashi, B. I., "Current Collection by a Long Conducting Cylinder in a Flowing Magnetized Plasma," Journal of Spacecraft and Rockets, Vol. 28, p. 592-598, September-October 1991.

Singh, R. B., "Three Dimensional Motion of a System of Two Cable-Connected Satellites in Orbit," Acta Astronautica, Vol. 18, No. 5, p. 301-308, 1973.

Stabekis, P., and Bainum, P. M., "Motion and Stability of Rotating Space Station-Cable-Counterweight Configuration," Journal of Spacecraft and Rockets, Vol. 7, No. 8, p. 912-918, 1970.

Stenzel, R.L. and Urrutia, J.M., "Currents Between Tethered Electrodes in a Magnetized Laboratory Plasma," Journal of Geophysical Research, Vol. 95, p. 6209-6226, May 1, 1990.

Stuiver, W., and Bainum, P. M., "A Study of Planar Deployment Control and Libration Damping of a Tethered Orbiting Interferometer Satellite," Journal of the Astronautical Sciences, Vol. 20, No. 6, p. 321-346, May-June 1973.

Stuiver, W., "Dynamics and Configuration Control of a Two-Body Satellite System," Journal of Spacecraft and Rockets, Vol. 11, No. 8, p. 345-346, 1974.

Sutton, G. W., and Diederich, F. W., "Synchronous Rotation of a Satellite at Less Than Synchronous Altitude," AIAA Journal, Vol. 5, No. 4, p. 813-815, 1967.

Swet, C. J., and Whisnant, J. M., "Deployment of a Tethered Orbiting Interferometer," Journal of the Astronautical Sciences, Vol. 17, No. 1, p. 44-59, July-Aug. 1969.

Tan, Z. and Bainum, P., M., "Optimal Linear Quadratic Gaussian Digital Control of an Orbiting Tethered Antenna/Reflector System," Journal of Guidance, Control and Dynamics, Vol. 17, No. 2, p. 234-241, April 1994.

Tiesenhausen, G. von, ed., "The Roles of Tethers on Space Station," NASA-TM-86519, NASA/MSFC, Oct. 1985.

Tyc, G., Vigneron, F.R. and Jablonski, A.M., "Two-Body Space Dynamics Technology Demonstration for the Biceps Small Satellite Mission," Canadian Aeronautics and Space Journal, Vol. 40, No. 1, p. 3-9, March 1994.

Tyc, G. and Han, R.P.S., "Attitude Dynamics Investigation of the OEDIPUS-A Tethered Rocket Payload," Journal of Spacecraft and Rockets, Vol. 32, No. 1, p. 133-141, February 1995.

Usui, H., Matsumoto, H. and Omura, Y., "Electron Beam Injection and Associated LHR Wave Excitation - Computer Experiments of Electrodynamic Tether System," Geophysical Research Letters, Vol. 18, p. 821-824, May 1991.

Usui, H., Matsumoto, H. and Omura, Y., "Plasma Response to High Potential Satellite in Electrodynamic Tether System," Journal of Geophysical Research, Vol. 98, No. A2, p. 1531-1544, February 1, 1993.

Vadali, S. R., "Feedback Tether Deployment and Retrieval," Journal of Guidance, Control and Dynamics, Vol. 14, p. 469-470, March-April 1991.

Vadali, S.R. and Kim, E.-S., "Feedback Control of Tethered Satellites Using Lyapunov Stability Theory," Journal of Guidance, Control and Dynamics, Vol. 14, p. 729-735, July-August 1991.

Vadali, S. R. and Kim, E., "Nonlinear Feedback Deployment and Retrieval of Tethered Satellite Systems," Journal of Guidance, Control and Dynamics, Vol. 15, p. 28-24, January-February 1992.

Vannaroni, G., Giovi, R. and DeVenuto, F., "Laboratory Simulation of the Interaction Between A Tethered Satellite System and the Ionosphere," Nuovo Cimento C, Serie 1, Vol. 15C, No. 5, p. 685-701, September-October 1992.

Vom Stein, R. and Neubauer, F.M., "Plasma Wave Field Generation by the Tethered Satellite System," Journal of Geophysical Research, Vol. 97, No. A7, p. 10,849-10,856, July 1, 1992.

von Flotow, A. H., and Williamson, P. R., "Deployment of a Tethered Satellite Pair into Low Earth Orbit for Plasma Diagnostics," The Journal of the Astronautical Sciences, Vol. 21, No. 1, p. 135-149, 1983.

Warnock, T. W. and Cochran, J. E., Jr., "Orbital Lifetime of Tethered Satellites," Journal of the Astronautical Sciences, Vol. 41, No. 2, p. 165-188, April-June 1993.

Wood, G. M., Siemers, P.M., Squires, R. K., Wolf, H. and Carlomagno, G. M., "Downward-Deployed Tethered Platforms for High-Enthalpy Aerothermodynamic

Research," Journal of Spacecraft and Rockets, Vol. 27, p. 215-221, March-April 1990.

Wright, A. N. and Schwartz, S. J., "The Equilibrium of a Conducting Body Embedded in a Flowing Plasma," Journal of Geophysical Research, Vol. 95, p. 4027-4038, April 1, 1990.

Yu, S., "On the Dynamics and Control of the Relative Motion Between Two Spacecraft," Acta Astronautica, Vol. 35, No. 6, p. 403-409, March 1995.

Zhu, R., Misra, A.K. and Modi, V.J., "Dynamics and Control of Coupled Orbital Motion of Tethered Satellite Systems," Journal of the Astronautical Sciences, Vol. 42, No. 3, p. 319-342, September 1994.

SECTION 8.0 CONTACTS

A -

Dr. Eduardo Ahedo
E.T.S.I. Aeronauticos
Plaza Cardenal Cisneros 3
28040 Madrid, SPAIN
3413366310

Mr. A.J. Alfonzo
Omitron, Inc.
6411 Ivy Lane
Suite 600
Greenbelt, MD 20770
301/474-1700

Mr. Andrew M. Allen
NASA, Johnson Space Center (CB)
Houston, TX 77058
713/244-8719

Prof. Yakov Alpert
Harvard Smithsonian Center
 for Astrophysics
60 Garden Street
Cambridge, MA 02138
617/495-7933

Dr. Jesus Pelaez Alvarez
E.T.S.I. Aeronauticos
Dpto. Fisica Aplicada
Pl. Cardenal Cisneros 3
28040, Madrid, SPAIN
3413366306

Mr. John Anderson
NASA Headquarters
Mail Code CC
Washington, DC 20546

202/358-4665

Prof. Francesco Angrilli
CISAS - University of Padova
Dept. of Mechanical Engineering
VIA Venezia, 1
35131, Padova, ITALY
39498286790

Mr. David A. Arnold
75 Woodbine Road
Belmont, MA 02178
617/484-7741

B -

Prof. Peter M. Bainum
Howard University
Dept. of Mechanical Engineering
2300 6th Street, N.W.
Washington, DC 20059
202/806-6612

Prof. Keith G. Balmain
University of Toronto
Dept. of Electrical and
  Computer Engineering
10 King's College Road
Toronto, Ontario, M5S 1A4 CANADA
416/978-3127

Mr. Ivan Bekey
NASA Headquarters
Advanced Concepts Office
Washington, DC 20546

Mr. Douglas Bentley
Cortland Cable Company
177 Port Watson Street
Cortland, NY 13045
607/753-8303

Prof. Silvio Bergamaschi
Padova University
Dept. of Mechanical Engineering
VIA Venezia 1
35131 Padova, ITALY
39498286809

Prof. Franco Bernelli-Zazzera
Politecnico di Ingegneria Aerospaziale
Politecnico di Milano
Via Golgi 40
20133 Milano, ITALY
39223994000

Dr. Thomas G. Berry
University of Manitoba
Dept. of Applied Mathematics
Winnipeg, Manitoba, R3T 2N2 CANADA
204/474-8345

Mr. Franco Bevilacqua
Alenia Spazio S.p.A. - Turin Plant
Advanced Studies Department
Corso Marche, 41
10146 Torino, ITALY
39117180718

Prof. Gianandrea Bianchini
University of Padova
Dept. of Mechanical Engineering
VIA Venezia 1
35131, Padova, ITALY
39498286808

Dr. Haik Biglari
Sverdrup Technology, Inc.
620 Discovery Drive
Huntsville, AL 35806
205/544-6890

Mr. Sven G. Bilen
University of Michigan
Space Physics Research Laboratory
2455 Hayward Street
Ann Arbor, MI 48109-2143
313/764-8461

Mr. Christopher Blunk
Eleanor Roosevelt High School
C/0 W.J. Webster, Jr.
NASA, Goddard Space Flight Center

244

Code 920.2
Greenbelt, MD 20771
301/286-4506

Dr. Carlo Bonifazi
Agenzia Spaziale Italiana
Viale Regina Margherita 202
00198 Roma, ITALY

Mr. Brian Briswell
Arizona State University
Dept. of Mechanical
 and Aerospace Engineering
Box 876106
Tempe, AZ 85287-6106
602/965-4363

Mr. Richard Brooke
12009 Lake Newport Road
Herndon, VA 22070

Mr. Larry L. Burgess
Lockheed Martin Astronautics
#3 Red Fox Lane
Littleton, CO 80127-5710
303/971-7521

Dr. William J. Burke
Phillips Laboratory/GPSG
29 Randolph Road
Hanscom AFB, MA 01731-3010
617/377-3980

C -

Mr. Michael A. Calabrese
NASA Headquarters
Code SS
300 E Street, S.W.
Washington, DC 20546
202/358-0899

Prof. Giovanni Carlomagno

Dip. di Energetica, Termofluidodinamica
Applicata e
Condizionamento Ambientale
Universita' di Napoli
P.le Tecchio, 80
80125 Napoli
39-81-7682178

Prof. Robert L. Carovillano
NASA Headquarters
Code SS
300 E. Street, S.W.
Washington, DC 20546
202/358-0894

Mr. Joseph A. Carroll
Tether Applications
1813 Gotham Street
Chula Vista, CA 91913-2624
619/421-2100

Dr. Kelly Chance
Harvard-Smithsonian Center
 for Astrophysics
60 Garden Street
MS 50
Cambridge, MA 02138
617/495-7389

Mr. Chia-Lie Chang
Science Applications International Corp.
1710 Goodridge Drive, T-2-3
McLean, VA 22102
703/734-5588

Dr. Franklin R. Chang-Diaz
NASA, Johnson Space Center
Houston, TX 77058
713/244-8923

Mr. Maurizio Cheli

NASA, Johnson Space Center
Houston, TX 77058
713/244-8739

Mr. Aaron Chilbert
Naval Research Laboratory
Code 8210
4555 Overlook Avenue, S.W.
Washington, DC 20375-5000

Dr. Palmer B. Chiu
NASA, Johnson Space Center
Automation, Robotics,
 and Simulation Division
NASA Road One; Mail Code ER6
Houston, TX 77058
713/483-8139

Dr. Dean Chlouber
System Planning Corporation
18100 Upper Bay Road, Suite 208
Houston, TX 77058
713/333-2666

Dr. Paul J. Coleman, Jr.
UCLA - IGPP
Inst of Geophysics & Planetary Physics
405 Hilgard Avenue
Los Angeles, CA 90095-1776
310/825-1776

Dr. Luis Conde
E.T.S.I. Aeronauticos
Dept. Fisica Aplicada
Pl. Cardenal Cisneros, 3
28040 Madrid, SPAIN
3413366305

Ms. Carolynn Conley
Muniz Engineering
P.O. Box 591672
Houston, TX 77259-1672
713/244-8150

Dr. David Cooke
Phillips Laboratory
PL/WSCF
Hanscom AFB, MA 01731-3010
617/377-2931

Dr. Mario L. Cosmo
Harvard-Smithsonian Center
for Astrophysics
60 Garden Street, MS 80
Cambridge, MA 02138
617/495-7412

Mr. Donald S. Crouch
Lockheed Martin
Mail Stop S8071
P.O. Box 179
Denver, CO 80201
303/977-3408

Mr. Kenneth H. Crumbly
3418 News Road
Williamsburg, VA 23188
757/258-5422

D -

Dr. Roberto Da Forno
CISAS - University of Padova
Via Venezia 1
35131 Padova, ITALY
39498286801

Mr. Mark A. Davis
7515 Mission Drive
Lanham, MD 20706
301/805-3960

Dr. Anthony DeCou
Northern Arizona University
College of Engineering
Box 5600
Flagstaff, AZ 86011
520/523-6114

Dr. Adarsh Deepak
Science and Technology Corporation
101 Research Drive
Hampton, VA 23666-1340
757/865-1894

Mr. John K. Diamond
NASA, Langley Research Center
M/S 471

Hampton, VA 23681-0001
757/864-1668

Prof. Luigi de Luca
Dip. di Energetica, Termofluidodinamica
Applicata e
Condizionamento Ambientale
Universita' di Napoli
P.le Tecchio, 80
80125 Napoli
39-81-7682182

Dr. Donald J. Dichmann
University of Maryland
Insitute for Physical Sciences
and Technology (IPST)
College Park, MD 20742
301/405-7887

Mr. Marino Dobrowolny
Instituto di Fisica dello Spazio
Interplanetario, CNR
00044 Frascati, ITALY
3969421017

Dr. Denis J. Donohue
Applied Physics Laboratory
Johns Hopkins University
Johns Hopkins Road
Laurel, MD 20723-6099
301/953-6258

Mr. Jean-Jacques Dordain
European Space Agency
8- 10 Rue Mario Nikis
75738 Paris Cedex 15, FRANCE
33153697338

Mrs. Patricia M. Doty
NASA, Marshall Space Flight Center
Mail Code FA64
Marshall Space Flight Center, AL 35812
205/544-4136

Mr. Michael Douglass
University of N Carolina - Chapel Hill
c/o W. J. Webster, Jr.
NASA/GSFC
Code 920.2
Greenbelt, MD 20771
301/286-4506

Dr. Adam T. Drobot
Science Applications International Corp.
1710 Goodridge Drive, T-2-3
McLean, VA 22102
703/734-5595

E -

Mr. Walter Eliuk
Bristol Aerospace Limited
660 Berry Street
P.O. Box 874
Winnipeg, Manitoba, R3C 2S4 CANADA
204/775-8331

Mr. Raymond A. Ernst
Lockheed Martin Astronautics
1725 Jefferson Davis Highway
Suite 300
Arlington, VA 22202
703/413-5762

Mr. Jay N. Estes
NASA, Johnson Space Center
Mail Code EG2

2101 NASA Road One
Houston, TX 77058
713/483-8379

Dr. Robert D. Estes
Harvard-Smithsonian Center
  for Astrophysics
60 Garden Street
Cambridge, MA 02138
617/495-7261

Dr. Steven W. Evans
NASA, Marshall Space Flight Center
Building 4203, Mail Stop EL58
Marshall Space Flight Center, AL 35812
205/544-8072

F -

Dr. Dale C. Ferguson
NASA, Lewis Research Center
Mail Stop 302-1
21000 Brookpark Road
Cleveland, OH 44135
216/433-2298

Mr. Enectali Figueroa
University of Puerto Rico
P.O. Box 5088 College Station
Mayaguez, 00681 PUERTO RICO
809/265-4846

Mr. Howard A. Flanders
Lockheed Martin Astronautics
Mail Stop S8071
P.O. Box 179
Denver, CO 80201
303/977-3669

Dr. Robert L. Forward
Tethers Unlimited

8114 Pebble Court
Clinton, WA 98236
360/579-1340

G -

Mr. Stephen Gates
Naval Research Laboratory
4555 Overlook Avenue, S.W.,
Code 8231
Washington, DC 20375
202/767-7680

Ms. Louise C. Gentile
Phillips Laboratory / GPSG
29 Randolph Road
Hanscom AFB, MA 01731-3010
617/377-7002

Dr. Sig Gerstl
Los Alamos National Lab
NIS/LDRD, MS-F658
Los Alamos, NM 87545
505/667-0952

Dr. Francesco Giani
Alenia Spazio S.p.A.
c. Marche 41
10146 Torino, ITALY
39117180716

Prof. Brian E. Gilchrist
University of Michigan
Space Physics Research Laboratory
2455 Hayward
Ann Arbor, MI 48109-2143
313/763-6230

Dr. John R. Glaese
Control Dynamics
Division of bd Systems
600 Boulevard South, Suite 304
Huntsville, AL 35803
205/882-2720

Mr. Howard Goldstein

NASA, Ames Research Center
Building 229-3
Moffett Field, CA 94030
415/604-6103

Dr. Michael A. Greenfield
HASA Headquarters
Office of Safety and Mission Assurance
300 E Street, S.W.
Washington, DC 20546
202/358-1930

Dr. Mario D. Grossi
Harvard-Smithsonian Center
 for Astrophysics
60 Garden Street
MS 80
Cambridge, MA 02138
617/495-7196

Dr. Umberto Guidoni
Italian Space Agency (ASI)
C/O NASA Johnson Space Center
Houston, TX 77058
713/244-2230

Dr. Gordon Gullahorn
Harvard-Smithsonian Ctr for Astrophysics
M/S 80
60 Garden Street
Cambridge, MA 02138
617/495-7419

H -

Ms. Linda Habash
University of Michigan
Space Physics Research Lab

2455 Hayward
Ann Arbor, MI 48109-2141
313/764-8461

Prof. Dr. -Ing W. Hallmann
Fachhochschule Aachen
Space Department
Hohenstaufenallee 6
D-52064 Aachen, GERMANY
24160092362

Prof. Ray P.S. Han
University of Manitoba
Dept. of Mechanical
 and Industrial Engineering
Winnipeg, Manitoba, R3T 2N2 CANADA
204/474-9519

Dr. David A. Hardy
Phillips Laboratory/GPSG
29 Randolph Road
Hanscom AFB, MA 01731-3010

Mr. James Harrison
NASA, Marshall Space Flight Center
Mail Code FA34
Huntsville, AL 35812
205/544-0629

Mr. Steven L. Hast
The Aerospace Corporation
P.O. Box 92957 M4/946
Los Angeles, CA 90009-2957
310/336-8968

Prof. Daniel E. Hastings
Massachusetts Institute of Technology
33-207 Department of Aero/Astro
77 Massachusetts Avenue
Cambridge, MA 02139
617/253-0906

Mr. Kazuo Ben Hayashida
NASA, Marshall Space Flight Center
ED52
Marshall Space Flight Center, AL 35812
205/544-4308

Prof. Roderick A. Heelis
Center for Space Sciences
Mail Stop: FO22
The University of Texas at Dallas
Box 830688
2601 N. Floyd Road
Richardson, TX 75080
214/883-2822

Mr. S. Herbiniere
CNES
18 aveneu Edouard Belin
31055 Toulouse Cedex, FRANCE
3361273439

MAJ Richard Higgins, Jr.
SMC/IMO
2420 Vela Way
Suite 1467-A5
Los Angeles AFB, CA 90245-4659
310/416-7651

Dr. J.M. Hinds
Space Telescope Science Institute
3700 San Martin Drive
Baltimore, MD 21218
410/338-4489

Dr. Noel W. Hinners
Lockheed Martin Astronautics
Mail Stop S-8000
P.O. Box 179
Denver, CO 80201
303/971-1581

Dr. Jeffrey A. Hoffman
NASA, Johnson Space Center
Mail Code CB
Houston, TX 77058
713/244-8723

Dr. John H. Hoffman
University of Texas at Dallas
2601 North Floyd Road
Richardson, TX 75080
214/883-2840

Dr. Toshihisa Honma
Massachusetts Institute of Technology
Dept. of Aeronautics and Astronautics
77 Massachusetts Avenue
Room 9-349
Cambridge, MA 02139-4307
617/258-7357

Mr. George D. Hopson
NASA, Marshall Space Flight Center
Marshall Space Flight Center, AL 35812
205/544-1735

Mr. Scott Horowitz
NASA, Johnson Space Center (CB)
Houston, TX 77058
713/244-8719

Dr. Robert P. Hoyt
Tethers Unlimited
8011 16th Avenue, N.E.
Seattle, WA 98115
206/525-9067

Mr. Brian Humphrey
Eleanor Roosevelt High School
C/O Dr. W.J. Webster, Jr.
NASA Goddard Space Flight Center

C/O 920.2
Greenbelt, MD 20771
301/286-4506

Prof. Franklin C. Hurlbut
University of California at Berkley
Dept. of Mechanical Engineering
6173 Etcheverry Hall
Berkley, CA 94720
510/642-7230

I -

Dr. Valerio Iafolla
C.N.R. I.F.S.I.
Via G. Galiles C.P.87
00044 Frascati RM, ITALY
39694186220

Dr. Devrie Intriligator
Carmel Research Center
P.O. Box 1723
Santa Monica, CA 90406
310/453-2983

Dr. George E. Ioup
University of New Orleans
Dept. of Physics
New Orleans, LA 70148
504/286-5591

Dr. Juliette W. Ioup
University of New Orleans
Dept. of Physics
New Orleans, LA 70148
504/286-6715

Mr. J.M. Gavira Izquierdo
ESTEC
YM Division
Keplerlaan 1
2200 AG Noordwijk, THE NETHERLANDS
31171984314

J -

Dr. Alexander M. Jablonski
Canadian Space Agency
DSM
6767 route de l'Aeroport
Saint-Hubert, Quebec, J3Y 8Y9 CANADA
514/926-4686

Dr. H. Gordon James
Communications Research Centre
P.O. Box 11490, Station "H"
Ottawa, Ontario, K2H 8S2 CANADA
613/998-2230

Mr. F.L. Janssens
ESTEC/ESA
Keplerlaan 1
Postbus 299
2200 AG Noordwijk, THE NETHERLANDS
31171983802

Mr. Les Johnson
NASA, Marshall Space Flight Center
Mial Code PS02
Huntsville, AL 35812
205/544-0614

Dr. R. Jerry Jost
System Planning Corporation
Center for Space Physics
18100 Upper Bay Road, Suite 208

Houston, TX 77058
713/333-2666

K -

Dr. Ira Katz
S-Cubed Division of Maxwell Labs
3398 Carmel Mountain Road
San Diego, CA 92121-1095
619/453-0060

Prof. Paul J. Kellogg
University of Minnesota
School of Physics and Astronomy
116 Church Street, S.E.
Minneapolis, MN 55455
612/624-1668

Dr. Vladomir Kim
Research Institute of Applied Mechanics
  and Electrodynamics of Moscow
  Aviation Institute,
4. Volokolam Snosse, Moscow, 125871,
RUSSIA
951580020

Dr. Kate P. Kirby
Harvard-Smithsonian Center
  for Astrophysics
60 Garden Street
MS 14
Cambridge, MA 02138
617/495-7237

Mr. Brian Kirouac
C/O Dr. W.J. Webster, Jr.
NASA, Goddard Space Flight Center
Code 920.2
Greenbelt, MD 20771
301/286-4506

Ms. Sheryl L. Kittredge

NASA, Marshall Space Flight Center
Mail Code ED63
Huntsville, AL 35812
205/544-9032

Mr. Stanislav l. Klimov
Space Research Institute of the
Russian Academy of Sciences
84/32 Profsoyuznaya
Moscow 117810, RUSSIA
70953331100

Mr. Joseph C. Kolecki
NASA, Lewis Research Center
Mail Stop 302-1
21000 Brookpark Road
Cleveland, OH 44135
216/433-2296

Mr. Paul Kolodziej
NASA, Ames Research Center
Mail Stop 234-1
Moffett Field, CA 94035
415/604-0356

Mr. William Kosmann
Interstel, Inc.
8000 Virginia Manor Road
Suite 180
Beltsville, MD 20705
301/210-0012

Mrs. Linda Habash Krause
University of Michigan
Space Research Building
2455 Hayward
Ann Arobr, MI 48109-2143
313/764-8461

Dr. Manfred Krishke
Kayser-Threde GmbH
80337 Munchen, GERMANY
498972495127

Mr. Frank M. Kustas
Lockheed Martin Astronautics
P.O. Box 179
Mail Stop F3085
Denver, CO 80201
303/971-9107

Mr. George Kyroudis
Spectrum Astro, Inc.
1440 N. Fiesta Boulevard
Gilbert, AZ 85234
602/892-8200

L -

Dr. J.G. Laframboise
York University
4700 Keele Street
North York, Ontario, M3J 1P3 CANADA
416/736-5621

Dr. James R. LaFrieda
The Aerospace Corporation
M6/210
P.O. Box 92957
Los Angeles, CA 90009
310/416-7177

Mr. David D. Lang
Lang Associates
2222 70th Avenue, S.E.
Mercer Island, WA 98040
206/236-2579

Dr. Jean-Pierre Lebreton
ESA/ESTEC

Mail Code 50
Kepleriann 1, 2200 AG Noordwijk,
THE NETHERLANDS
31171983600

Mr. Enzo Letico
ASI Washington Representative
Italian Space Agency
250 E. Street, S.W., Suite 300
Washington, DC 20024
202/863-1298

Mr. George M. Levin
NASA Headqaurters
Mail Code MP
300 E Street, S.W.
Washington, DC 20546-0001
202/358-4478

Dr. Yevgeniy M. Levin
University of Minnesota
Dept. of Aerospace Engineering
 and Mechanics
107 Akerman Hall
Minneapolis, MN 55455
612/933-3796

Dr. Mark J. Lewis
University of Maryland
Dept. of Aerospace Engineering
College Park, MD 20742-3015
301/405-1133

Dr. Weiwei Li
York University
Dept. of Physics and Astronomy
4700 Keele Street
North York, Ontario, M3J 1P3 CANADA
416/736-2100

Dr. Renato Licata

Alenia Spazio S.p.A. - Turin Plant
Control & Dynamics Department
Corso Marche, 41
10146 Torino, ITALY
39117180233

Dr. Garry M. Lindberg
Canadian Space Agency
6767 Route de L'Aeroport
St-Hubert, Quebec, J3Y 8Y9 CANADA
514/926-4372

Mr. C.R. Lippincott
University of Texas at Dallas
William B. Hanson Center for
  Space Sciences
P.O. Box 830688 / M/SF022
Richardson, TX 75083-0688
214/883-2819

Prof. James M. Longuski
School of Aeronautics &Astronautics
Purdue University
1282 Grissom Hall
West Lafayette, IN 47907-1282
317/494-5139

Dr. Enrico Lorenzini
Harvard-Smithsonian Centerrophysics
  for Astrophysics
60 Garden Street
MS 80
Cambridge, MA 02138
617/495-7211

Dr. Charles A. Lundquist
The University of Alabama
  in Huntsville
301 Sparkman Drive
Huntsville, AL 35899
205/895-6620

Dr. Andrea A.E. Luttgen

University of Toronto
Dept. of Elec. & Comp. Engineering
10 King's College Road
Toronto, Ontario, M5S 1A4 CANADA
416/978-5831

M -

Mr. Bruce A. Mackenzie
Draper Laboratory
M/S 22
555 Technology Square
Cambridge, MA 02139
617/258-2828

Mr. Robert J. Mahoney
NASA, Johnson Space Center
Mail Code DT23 / Rendezfous Training
Houston, TX 77058
713/244-7377

Mr. Franco Malerba
Via Contore 10/6
Genova 16149, ITALY
39106450365

Dr. Gianfranco Manarini
Agenzia Spaziale Italiana
Viale Regina Margherita 202
00198 Rome, ITALY
3968567361

Prof. Franco Mariani
Via Ricerca Scientifica 11
00133 Roma, Italy
39678792319

Dr. Hartmut Marschall
Universitat Koln
Institut f. Geophysik und Meteorologie
Albertus-Magnus-Platz
Koln 50923, GERMANY

255

492214703387

Mr. Leland S. Marshall
Lockheed Martin Astronautics
P.O. Box 179
Mail Stop 58071
Denver, CO 80201
205/544-1927

Mr. Patrick Martin
Vitro Corporation
400 Virginia Avenue
Washington, DC
202/646-6371

Mr. Manuel Martinez-Sanchez
Massachusetts Institute of Technology
Dept. of Aeronautics and Astronautics
Cambridge, MA 02139
617/253-5613

Mr. Robert O. McBrayer
NASA, Marshall Space Flight Center
Code JA71
Marsahll Space Flight Center, AL 35812
205/544-1926

Dr. James E. McCoy
NASA, Johnson Space Center
Code 5N3
Houston, TX 77058
713/483-5068

Dr. Leonard T. Melfi
Science and Technology Corporation
101 Research Drive
Hampton, VA 23666-1340
757/865-1894

Dr. S.B. Mende
Dept. 91-20, 252
Lockheed

3251 Hanover St.
Palo Alto, CA 94024

Dr. Pietro Merlina
Alenia Spazio S.p.A. - Turin Plant
Advanced Studies Department
Corso Marche, 41
10146 Torino, ITALY
39117180718

Dr. Luca Minna
Advanced Engineering Technology
Corte Lambruschini - Piazza Borgo Pila 40
16129 Genova - ITALY
39105531425

Prof. Arun K. Misra
McGill University
Dept. of Mechanical Engineering
817 Sherbrooke Street West
Montreal, CQ, H3A 2K6 CANADA
415/398-6288

Prof. Antonio Moccia
Dip. di Scienza e Ingegneria dello Spazio,
Univ. di Napoli, P.le Tecchio 80, 80125 Napoli,
Italy,
39-81-7682158

Dr. Vinod J. Modi
University of British Columbia
Dept. of Mechanical Engineering
2324 Main Mall
Vancouver, B.C., V6T 1Z4 CANADA
604/822-2914

Mr. Richard Moyer
Advent Systems, Inc.
P.O. Box 222861
Chantilly, VA 22021
703/631-3498

Mr. Ronald M. Muller
NASA, Goddard Space Flight Center
Code 170
Mission to Planet Earth Office
Greenbelt, MD 20771
301/286-9695

Ing. Bruno Musetti
Alenia Spazio S.p.A. - Turin Plant
Control and Dynamics Department
Gruppo Sistemi Spaziali
Corso Marche, 41
10146 Torino, ITALY
39117180752

N -

Mr. Claude Nicollier
NASA, Johnson Space Center
Houston, TX 77058

Ms. Penny L. Niles
Lockheed Martin Astronautics
P.O. Box 179
M/S SB071
Denver, CO 30201
303/977-3679

Mr. Stephen T. Noble
Rice University
P.O. Box 244
Tavernier, FL 33070
305/852-8879

Mr. Mauro Novara

ESA/ESTEC
Postbus 299
2200 AG Noordwijk, THE NETHERLANDS
31171984003

O -

Mr. Marvin L. Odefey
Lockheed Martin Astronautics
P.O. Box 179
Mail Stop S8110
Denver, CO 80201
303/977-7782

Mrs. Cinthya Ottonello
University of Genoa
Dipartimento di Ingegneria Biofisica
 ed Elettronica
Via all' Opera Pia 11 A
16145, Genova, ITALY
39103532187

Dr. K.-I. Oyama
The Institute of Space and Astronautical Science
3-1-1, Yoshinodai, Sagamihara, Kanagawa 229
Japan
0427513911

P -

Mr. L. Marco Palenzona
ESA/ESTEC

Postbus 299
2200 AG Noordwijk (zH),
THE NETHERLANDS
31171983651

Dr. Dennis Papadopoulos
University of Maryland
Dept. of Physics
College Park, MD 20742
301/405-1526

Dr. Monica Pasca
Universita di Roma "La Sapeinza"
Dipartimento di Ingegneria
 Strutturale e Geotecnica
Via Eusossiana, 18-00184 Roma, ITALY
39644585156

Mr. Barry R. Payne
Bristol Aerospace Limited
600 Berry Street
Winnipeg, Manitoba, R3C 2S4 CANADA
204/775-8331

Mrs. Amey R. Peltzer
Naval Research Laboratory
Code 8123, Building 58 R127
4555 Overlook Avenue, S.W.
Washington, DC 20375
202/767-3982

Dr. Paul A. Penzo
Jet Propulsion Lab
1800 Oak Grove Drive
Pasadena, CA 91109
818/354-6162

Mrs. Maria Antonietta Perino
Alenia Spazio S.p.A.
Corso Marche, 41
10146 Torino, ITALY
39117180712

Mr. Paolo Piantella
Alenia Spazio S.p.A.
Corso Marche, 41
10146 Torino, ITALY
3911712923

Mr. Engelbert Plescher
FH Aachen, Space Department
Hohenstaufenallee 6
D-52064, Aachen, GERMANY
24160092394

Prof. J. David Powell
Stanford University
Aero/Astro Department
MC 4035
Stanford, CA 94305
415/723-3425

Prof. Jordi Puig-Suari
Arizona State University
Dept. of Mechanical
  and Aerospace Engineering
Box 876106
Tempe, AZ 85287-6106
602/965-4363

R -

Prof. W. John Raitt
Utah State University
Dept. of Physics
CASS/UMC 4405
Logan, UT 84322
801/797-2849

Mr. Ray D. Rhew
NASA, Langley Research Center
M/S 238
Hampton, VA 23681
757/864-4705

Mr. Charles C. Rupp
NASA, Marshall Space Flight Center
Mail Code PS04
Huntsville, AL 35812
205/544-0627

S -

Mr. Dieter Sabath
Lehrstuhl fur Raumfahrtechnik, TU
Munchen, GERMANY
498921052176

Mr. Jean Sabbagh
Agenzia Spaziale Italiana
V.le Regina Margherita, 202
00198 Roma, ITALY

Ms. Salma I. Saeed
Stanford University
P.O. Box 14213
Stanford, CA 94309
415/725-3297

Prof. Juan R. Sanmartin
Escuela Tecnica Superior de
  Ingenieros Aeronauticos
Universidad Politecnica de Madrid
Plaza Cardenal Cisneros, 3
28040 Madrid, SPAIN
3413366302

Mr. Andrew D. Santangelo
The Michigan Technic Corporation
17133 Inavale
Holland, MI 49424
616/399-4045

Dr. S. Sasaki
The Institute of Space and Astronautical Science
3-1-1, Yoshinodai, Sagamihara, Kanagawa 229
Japan
0427513911

Mr. Chikatoshi Satoh
Nihon University
College of Science and Technology
Dept. of Aerospace Engineering
7-24-1, Narashinodai, Funabashi-City,
274 JAPAN

Mr. Andreas Schroeer
Ruhr-Universiteet Bochum
Theoretische Physik IV
D-44780 Bochum, GERMANY
492347003729

Mr. John D. Schumacher
NASA Headquarters
Acting Assoicate Administrator
External Affairs
Washington, DC 20546

Mr. Thomas J. Settecerri
Lockheed Martin Astronautics
2400 NASA Road 1 / C-104
Houston, TX 77058
713/483-4160

Prof. Irwin I. Shapiro
Harvard-Smithsonian Center
  for Astrophysics
60 Garden Street
MS 45
Cambridge, MA 02138

Dr. Carl L. Siefring
Naval Research Laboratory
Plasma Physics Division
Code 6755
Washington, DC 20375
202/767-2467

Mr. Charles W. Shaw
NASA, Johnson Space Center
Flight Director Office
Mail Code DA8
Houston, TX 77058
713/483-5416

Mr. H. Frayne Smith
NASA, Marshall Space Flight Center
Mail Code EJ23
Huntsville, AL 35812
205/544-3572

Dr. Roberto Somma
Alenia Spazio S.p.A.
via Saccomuro, 24
00131 - Rome, ITALY
39641513208

Ms. Becky C. Soutullo
NASA, Marshall Space Flight Center
TSS Project Office
MSFC/JA71
Marshall Space Flight Center, AL 35812
205/544-1977

Mr. David Spencer
Naval Research Laboratory
Code 8213
4555 Overlook Avenue, S.W.
Washington, DC 20375-5000
202/767-6425

Dr. Curtis H. Spenny
U.S. Air Force Institute of Technology
AFIT/ENY
2950 P Street
Building 140
Wright-Patterson AFB, OH 45433-7765
513/753-6565

Mr. John H. Stadler
NASA, Langley Research Center
Mail Stop 431
Hampton, VA 23681-0001
757/864-7076

Dr. Wolfgang Steiner
Technical University of Vienna

Institute of Mechanics
Wiedner Hauptstrasse 8-10/325
A-1040 Vienna, AUSTRIA
431588015516

Dr. Nobie H. Stone
NASA, Marshall Space Flight Center
Space Physics Laboratory
Huntsville, AL 25812
205/544-7642

Ing. Bruno Strim
Alenia Spazio S.p.A. - Turin Plant
Scientific Satellites Directorate
Corso Marche, 41
10146 Torino, ITALY
39117180733

Mr. Locke M. Stuart
NASA, Goddard Space Flight Center
Code 920
Greenbelt, MD 20771
301/286-6481

Mr. Thomas D. Stuart
NASA Headquarters
Mail Code MO
Washington, DC 20560
202/358-4422

Mr. Francesco Svelto
ASI
Viale Regina Margherita 202
V.Le Rue Margerita, 202
00198 Rome, ITALY

T -

Prof. Giorglo Tacconi
Universita' di Genova
Dipartimento Ingegneria Biofisica
 ed Elettronicat
Via Opera Pia, 11 A
Genova, 16145, ITALY
39103532187

Dr. David L. Talent
Lockheed Martin Astronautics
2400 NASA Road 1
Mail Code C-104
Houston, TX 77058
713/483-5837

Mr. Zhaozhi Tan
Howard University
2300 6th Street, N.W.
Washington, DC 20059
202/806-4842

Dr. Patrick T. Taylor
NASA, Goddard Space Flight Center
Geodynamics
Code 921
Greenbelt, MD 20771
301/286-5412

Mr. Paul Tetzlaff
Daimler-Benz Aerospace / RST Rostock
Am Strom 109
D-18119 Warnemunde, GERMANY
4938156259

Ms. Shannon Thornburg
Stanford University
419 E. Indiana Street
Rapid City, SD 57701
605/394-2452

Mr. Donald D. Tomlin
NASA, Marshall Space Flight Center
Structures and Dynamics Lab
Mail Code ED13
Marshall Space Flight Center, AL 35812
205/544-1465

Mr. P. Chewning Toulmin
Hughes - STX
C/O Dr. W.J. Webster, Jr.
NASA, Goddard Space Flight Center
Code 920.2
Greenbelt, MD 20771
301/286-4506

Mr. Steven G. Tragesser

Purdue University
1282 Grissom Hall
West Lafayette, IN 47707-1282
317/494-5813

Prof. Hans Troger
Technical University of Vienna
Institute of Mechanics
Wiedner Hauptstrasse 8-10/325
A-1040 Vienna, AUSTRIA
431588015510

Mr. George Tyc
Bristol Aerospace Limited
660 Berry Street
P.O. Box 874
Winnipeg, Manitoba, R3C 2S4 CANADA
204/775-8331

U -

Dr. Maximilian Ullmann
Fiat
1776 Eye Street, N.W. #775
Washington, DC 20006
202/862-1614

V -

Prof. Sergio Vetrella
Dip. di Ingegneria Aerospaziale
Seconda Universita' di Napoli
Via Roma, 29
81031 Aversa (CE)
tel +39-81-5044035

Dr. Frank R. Vigneron
Canadian Space Agency
P.O. Box 11490
Station H

Ottawa, Ontario, K2H 82S CANADA
613/998-2741

Mr. Giuseppe Viriglio
Alenia Spazio S.p.A.
Corso Marche 41
10146 Torino, ITALY

W -

Mr. Bruce K. Wallace
NASA, Marshall Space Flight Center
EL64
Marshall Space Flight Center, AL 35812
205/544-1306

Dr. William J. Webster, Jr.
NASA, Goddard Space Flight Center
Code 920.0
Greenbelt, MD 20771
301/286-4506

Mr. Kenneth J. Welzyn
NASA, Marshall Space Flight Center
Flight Dynamics Branch
ED-13
Marshall Space Flight Center, AL 35812
205/544-1731

Mr. Scott L. Wetzel
Allied Signal Technical Services Corp.
NASA SLR
7515 Mission Drive
Lanham, MD 20706
301/805-3987

Prof. Paul J. Wilbur
Colorado State Univesity
Dept. of Mechanical Engineering
60 Garden Street
Fort Collins, CO 80523
303/491-8564

Mr. John Williams
Hughes Research Laboratory
3011 Malibu Canyon Road
Malibu, CA 90265
310/317-5446

Mr. Scott Williams
Stanford University
STARIAB/ Durand 33 / 4055
Stanford, CA 94305
415/725-0482

Dr. Thomas L. Wilson
NASA, Johnson Space Center
Houston, TX 77058
713/483-2147

Mr. Dennis Ray Wingo
University of Alabama in Huntsville
Center for Space Plasama
 and Aeronomic Research
Huntsville, AL 35899
205/895-6912

Dr. George L. Withbroe
NASA Headquarters
Code SS
Washington, DC 20546
202/358-1544

Dr. George M. Wood
101 Research Drive
Hampton, VA 23666-1340
757/865-1894

Mr. Alfred C. Wright
Lockheed Martin Astronautics
2550 E. Pine Bluff Lane
Highlands Ranch, CO 80210
303/977-5952

Mr. N. Convers Wyeth
Science Applications International Corp.
1710 Goodridge Drive
M/S 2-3-1
McLean, VA 22102
703/821-4411

Mr. Michael F. Zedd
Naval Research Laboratory
4555 Overlook Avenue, S.W.
Code 8233
Washington, DC 20375-5355
202/404-8337

Z -

Acknowledgments

This edition of the handbook is dedicated to the memory of the people involved in the advancement of tethers who are no longer with us, among them Stanley Shawan, Billy Nunley and Silvio Bergamaschi.